Managing Chemical Safety

Dennis G. Nelson

ABS Consulting

Government Institutes
Rockville, MD

ABS Consulting

Government Institutes
4 Research Place, Rockville, Maryland 20850, USA
Phone: (301) 921-2300
Fax: (301) 921-0373
Email: giinfo@govinst.com
Internet: http://www.govinst.com

Library of Congress Cataloging-in-Publication Data

Nelson, Dennis (Dennis Glenn)
    Managing chemical safety / Dennis Nelson.
      p. cm.
    Includes bibliographical references and index.
    ISBN 0-86587-838-2
    1. Chemical engineering--Safety measures. I. Title.

    TP150.S24 N45 2002
    660'.2804--dc21

                             2002066484

Contents

Chapter 2
Fires and Explosions

Chapter 3
Chemical Process Design Considerations

Chapter 4
Accident and Incident Investigation

Chapter 5
Safe Work Conditions

Chapter 6
Safe Work Practices

Chapter 7
Accountability and Performance Measures

Chapter 8
Regulatory Requirements

Chapter 9
Safety and Health Training

Chapter 10
Implementing a Chemical Safety
Management System

Chapter 11
Developing a Winning Game Plan

Appendix A
Using EPA's *Risk Management Program Guidance*

Appendix B
Modeling Software

List of Figures and Tables

Preface

Chemical Safety Management is written for safety managers and coordinators. The book is appropriate for facility and staff safety personnel as well as technical professionals. Directors, managers, and first-line supervisors will benefit from the "how to" that is needed to comply with government regulations, formulate internal policies, and adopt safe work practices in any workplace in which chemicals are present. The chapters should appeal to those with and without a background in chemistry.

There are huge benefits to chemical safety. Based on an eight-year study of major accidents associated with hazardous materials, OSHA estimated that more than 300 fatalities and nearly 2000 injuries and illnesses per year would be avoided through full compliance with Process Safety Management.

Yet implementing a chemical safety management system takes time and dedication to continuous improvement. Readers of this book need to help their organizations turn compliance into a management tool. Policies and procedures must be in place so that everyone can work together to bring chemical safety to the point where it is "habit strength."

Safety is a condition of employment for all of us. We are all responsible to work safely and to be a model for other employees. Safety is also a line-management responsibility. Safety and health managers are a vital resource, but the entire organization must work together to create a safe work environment. And by utilizing the strategies presented in this book, hopefully any employee can find that safety is as manageable as anything else that goes on in the organization.

Introduction

Chemical safety management is about much more than meeting legal requirements. Safety, health, and a concern for environmental stewardship must be an integral part of every facility's program. Most safety professionals can tell you *why* accidents happen. Although this book stresses the importance of discovering the root cause(s) of accidents, this is not a "why" book. This is a "how to" book. How does one go from the current starting point to effective chemical safety management?

A Preview of What's Ahead

Chapter 1 addresses toxic hazards. Facilities that process and otherwise handle a variety of inorganic and organic chemicals need to have access to people who are knowledgeable in this topic.

Examples of toxic chemicals and their primary routes into the body are discussed, as well as material safety data sheets (MSDSs), acute and chronic exposures, and chemical exposure limits set by OSHA and the American Conference of Governmental Industrial Hygienists (ACGIH).

The safety precedence sequence is introduced as a tool for setting hazard elimination and control priorities. Chemical hazard control strategies and techniques are demonstrated that range from chemical substitution to use of personal protective equipment (PPE).

This chapter also addresses toxic release source and dispersion models. It is important to model major and limited release sources in order to predict the potential on and off-site impact of a release. The reader will recognize the need to control hazards at their source rather than far downstream in the process, at the "end of the pipe." Finally, the chapter concludes with preventive and protective mitigation methods to lower the risk of a chemical release incident.

Chapter 2 examines fires, explosions, and other emergencies. Specific topics include fire and explosion basics, classification of small fires and fire extinguishers, properties of liquid, gaseous and solid (dust) fuels;

controlling ignition sources, fire protection and detection, emergency squads, as well as fire training and drills. The chapter concentrates on control of flammable liquids and gases and how to find specification information for storage and handling of these common industrial materials.

Emergency planning and response is the final section of Chapter 2. In setting up an overall facility Environmental, Health, and Safety (EHS) program, it is important to address potential security issues and to plan/practice for the types of emergencies the facility is capable of handling.

Chapter 3 examines chemical process design considerations that have enormous "up-front" leverage on safety and health. Design decisions made at the initial stage of an engineering project have a great impact on personnel and process safety in the plant. Companies need to "design-in" layers of protection that will minimize chemical exposure to employees and the public while facilities and equipment are in the conceptual stage of development.

The chapter stresses the importance of including process flow diagrams, material and energy balances, piping and instrumentation diagrams, and other chemical, process, and equipment information into a total design package. The engineering project process is also examined, as well as the use of design and hazard reviews in integrating safety into the eight stages of a project. The chapter discusses design strategies for safety and provides examples of design considerations.

The chapter concludes with key points for design of chemical process systems.

Chapter 4 covers accident and incident investigation. A thorough investigation leads to important activities that prevent future incidents of a similar nature and have significant financial benefits as well.

This chapter demonstrates how critical it is to pre-plan and install an *accident-incident investigation process.* The chapter discusses investigation or analysis procedures for potential incidents that meet OSHA's process safety management (PSM) requirements. Finally, Chapter 4 provides examples of management systems and tools that will help the reader establish their own facility accident-incident investigation system.

Chapter 5 helps the chemical safety manager establish and maintain safe work conditions in the facility through a discussion of systems and tools used to prevent major and other accidents/incidents, near accidents, and potential hazards associated with unsafe work conditions.

The chapter provides a simple inspection procedure. It also provides tools for workplace self-inspections and follow-up as well as methods to assure safe work conditions related to process safety and risk management.

Chapter 6, "Safe Work Practices," complements Chapter 5, "Safe Work Conditions." This chapter examines tools such as job safety analysis to help safety coordinators and supervisors establish safe work procedures. It also provides a framework and examples of behavioral safety systems applied to chemical safety management.

Chapter 7 addresses the link between accountability and measurement of safety. In addition to demonstrating the accountability-measurement link, the chapter provides examples of positive measures of safety performance. It distinguishes between results measures that are outcomes of non-conformance and measures of upstream activities (performance indicators) that will result in continuous, incremental safety improvement. Finally, the chapter discusses examples of traditional results measures and performance indicators that relate to improving chemical safety management systems.

Chapter 8 provides the reader with a regulatory perspective on safety, health, and environmental requirements. An outline of the basic OSHA, EPA, and DOT standards that relate to chemical safety management is included. The reader will gain an understanding of regulatory requirements for process safety management (PSM), hazard communication or employee right-to-know, hazardous waste operations and emergency response (HAZWOPER), risk management planning (RMP), and the labeling-transport of hazardous materials.

Finally, the chapter provides a structure of safety, health, and environmental standards on which to build company and facility policies and procedures.

Chapter 9 discusses of safety and health training. The primary goal of the chapter is to demonstrate the importance of integrating top-priority product-process and chemical safety, health, and environmental training. This chapter describes a basic learning model, and provides the reader with procedures related to chemical operator performance. The chapter also compares conventional and multi-media training systems in terms of capabilities, advantages, and limitations. Finally, it provides a discussion and an example of a training matrix used to schedule specific safety and health training topics.

Chapter 10 provides a framework for program implementation. It provides nine elements of a chemical safety management system that can be installed in a parallel fashion.

Included in the steps is an explanation of the importance of company and facility management's vision for how a fully installed chemical safety management system would operate in order to set effective goals, objectives, and performance measures. The chapter also stresses the need for an annual plan to establish measurable objectives and to use as a basis for gauging accomplishment.

Hazard-risk evaluation and control systems are central to all facility safety programs. This chapter demonstrates how to link regulatory requirements for programs like process safety management and hazard communication with policies, procedures, practices, and systems/tools for hazard evaluation and control.

The final section of Chapter 10 is devoted to installing an EHS management information system (MIS) that will meet the needs of the plant and company as well as a diverse group of stakeholders.

Chapter 11 provides strategies and tips for developing successful chemical safety management systems. Two assessment methods are offered as the basis for installing a driving force for change. Finally, the chapter provides a discussion of how to integrate the information provided in each of the chapters into a cohesive management system.

Toxic Hazards

Introduction

It is important for facility safety and health managers to apply the basic principles of toxicology in order to make more accurate and complete workplace assessments. The manager needs to quantify exposures that are occurring, and should work with manufacturing and plant management to determine what to do about these hazards. This can be broken down into three steps: *hazard identification, hazard evaluation,* and *hazard control.*

Facility medical staff and safety/health managers need to be familiar with the toxicity of their chemical inventory, including acute and chronic exposure effects, and reversible and irreversible effects. The purpose of this chapter is to provide the readers with a basic background in toxic chemicals found in the workplace. Safety professionals need to take into account the effect chemicals have on the body and the permissible exposure limits as defined by professional and regulatory groups.

Toxic Hazard Identification

Types of Toxic Hazards

Safety managers need to be aware that toxic hazards exist in many forms, including:

- Chemicals: solids, liquids, gases, and vapors
- Dusts and fibers
- Noise
- Radiation

All toxic hazards can be classified into two categories: acute hazards and chronic hazards. Acute exposure hazards are exposures that are severe but last for a short period of time; usually, the effects of acute exposures are more reversible than the effects of chronic exposures. An example of an acute hazard would be an employee sprayed in the face with anhydrous ammonia during an improperly conducted line-breaking procedure.

Chronic exposure hazards occur when someone is repeatedly exposed to a toxic substance over a (longer) period of time. The effects of this type of exposure usually become evident only after multiple exposures. For example, repeated exposure to toluene causes dry, irritated skin; redness; and respiratory, liver, and central nervous system damage.

Routes of Entry into the Body and Effects

Exposure to toxicants occurs through:

- Contact with skin
- Ingestion
- Inhalation
- Contact with eyes
- Cuts and punctures

There are two types of effects: reversible and irreversible effects. A reversible effect is one in which the condition resulting from the exposure (for example, dry skin) is restored to the original state (healthy skin). An irreversible effect is one where the original condition cannot be restored (for example, repeated exposure to toluene diisocyanate).

Material Safety Data Sheets

Describing the hazards of specific chemicals that could enter the body by inhalation, through the skin and eyes, and by mouth is an OSHA requirement. This information must be provided by the chemical's manufacturer or importer by means of a "hazard determination" of toxicological effects on the body.[1] OSHA does not require the plant using the chemicals to make this determination unless they choose not to accept information provided by the manufacturer/importer. The required information must be provided to the user in the form of a material safety data sheet (MSDS), required by section (g) of the Hazard Communication Standard. The MSDS format and volume of information has evolved considerably over the years, and most companies use their own format. However, it is best to follow the basic outline required by OSHA's eight sections or the European Union's 16 sections.

By comparing different sources of MSDS information on a chemical, the reader will discover that the quantity and quality of information varies considerably from source to source. For this reason, many companies direct their industrial hygiene department to summarize safety and health information—for example in a *Chemical Safety Information Summary (CSIS)*. Despite the differences, MSDSs are the most important source of chemical health and safety information for most facilities.

Figure 1-1 is a good example of a modern MSDS (updated on a regular basis) that provides safety and health information for anhydrous ammonia.

Product: *Ammonia, Anhydrous*

Form No.: P-4562-E   Date: October 1997

1. Chemical Product and Company Identification

Product Name: Ammonia, Anhydrous (MSDS No. P-4562-E)
Trade Name: Ammonia, Anhydrous
Chemical Name: Ammonia
Synonyms: Ammonia Gas, Spirit of Hartshorn
Formula: NH_3
Chemical Family: Alkaline Gas
Telephone:
  Emergencies: 1-800-645-4633*
  CHEMTREC: 1-800-424-9300
  Routine: 1-800-PRAXAIR
Company Name: Praxair, Inc.
    39 Old Ridgebury Road
    Danbury, CT 06810-5113

*Call emergency number 24 hours a day only for *spills, leaks, fire, exposure, or accidents involving this product*. For routine information contact your supplier, Praxair sales representative, or call 1-800-PRAXAIR (1-800-772-9247)

2. Composition / Information on Ingredients

For custom mixtures of this product, request a Material Safety Data Sheet for each component. See Section 16 for important information about mixtures.

Ingredient Name	CAS Number	Percentage	OSHA PEL	ACGIH TLV
Ammonia, Anhydrous	7664-41-7	>99%*	50 ppm	25 ppm (TLV-STEL, 15 min=35 ppm)

The symbol ">" means "greater than."

Figure 1-1 Praxair® Material Safety Data Sheet For Anhydrous Ammonia[2]

Product: Ammonia, Anhydrous   Form No.: P-4562-E   Date: October 1997

3. Hazards Identification

EMERGENCY OVERVIEW
DANGER! Corrosive liquid and gas under pressure.
Harmful if inhaled.
Causes eye, skin, and respiratory tract burns.
May cause kidney and respiratory system damage
Self-contained breathing apparatus
must be worn by rescue workers.
Odor: Pungent, irritating

THRESHOLD LIMIT VALUE: TLV-TWA = 25 ppm, TLV-STEL (15 min) = 35 ppm (ACGIH 1997). TLV-TWA should be used as a guide in the control of health hazards and not as fine lines between safe and dangerous concentrations.

EFFECTS OF A SINGLE (ACUTE) OVEREXPOSURE:
INHALATION—Overexposure to concentrations moderately above the Threshold Limit Value (TLV) of 25 ppm may irritate the eyes, nose, and throat. Higher concentrations may cause breathing difficulty, chest pain, bronchospasm, pink frothy sputum, and pulmonary edema. Overexposure may predispose to acute bronchitis and pneumonia.
SKIN CONTACT—Liquid may cause moderate to severe redness, swelling, and ulceration of the skin, depending on the degree and duration of contact. At high concentrations, gas may cause chemical burns. Prolonged or widespread skin contact may result in the absorption of potentially harmful amounts of material.
SWALLOWING—An unlikely route of exposure; this product is a gas at normal temperature and pressure. But exposure, should it occur, may cause chemical burns of the mouth, throat, esophagus, and stomach.
EYE CONTACT—Liquid may cause pain, severe redness, and swelling of the conjunctiva, damage to the iris, corneal opacification, glaucoma, and cataracts. Gas may cause pain and excessive tearing with acute corneal injury at high concentrations.

EFFECTS OF REPEATED (CHRONIC) OVEREXPOSURE: Chronic exposure may cause chemical pneumonitis and kidney damage.
OTHER EFFECTS OF OVEREXPOSURE: None known.
MEDICAL CONDITIONS AGGRAVATED BY OVEREXPOSURE: Inhalation may aggravate asthma and inflammatory or fibrotic pulmonary disease. Skin irritation may aggravate an existing dermatitis.
SIGNIFICANT LABORATORY DATA WITH POSSIBLE RELEVANCE TO HUMAN HEALTH HAZARD EVALUATION: None known.
CARCINOGENICITY: Ammonia, Anhydrous is not listed by NTP, OSHA, and IARC.

Figure 1-1 Praxair® Material Safety Data Sheet For Anhydrous Ammonia (cont.)

Product: Ammonia, Anhydrous   Form No.: P-4562-E   Date: October 1997

4. First Aid Measures

INHALATION: Remove to fresh air. If not breathing, give artificial respiration. **WARNING: Rescuer may receive chemical burns from giving mouth-to-mouth resuscitation.** If breathing is difficult, qualified personnel may give oxygen. Keep patient warm. Call a physician.

SKIN CONTACT: Immediately flush skin with plenty of water for at least 15 minutes while removing contaminated clothing and shoes. Discard clothing and shoes. Call a physician.

SWALLOWING: An unlikely route of exposure; this product is a gas at normal temperature and pressure. Give at least two glasses of water or milk at once. Do not induce vomiting. Call a physician.

EYE CONTACT: Immediately flush eyes thoroughly with water for at least 15 minutes. Hold the eyelids open and away from the eyeballs to ensure that all surfaces are flushed thoroughly. See a physician, preferably an ophthalmologist, immediately.

NOTES TO PHYSICIAN: *Victims of overexposure should be observed for at least 72 hours for delayed edema. The hazards of this material are mainly due to its severe irritant and corrosive properties on the skin and mucosal surfaces. There is no specific antidote, and treatment should be directed at the control of symptoms and clinical condition.*

5. Fire Fighting Measures

FLASH POINT (test method): Flammable Gas
AUTOIGNITION TEMPERATURE: 1204 deg. F (651 deg. C)
FLAMMABLE LIMITS IN AIR (% by volume):
LOWER 15%
UPPER 28%
EXTINGUISHING MEDIA: CO_2, dry chemical, water spray, or fog.
SPECIAL FIRE FIGHTING PROCEDURES: DANGER! Toxic, corrosive, flammable liquefied gas under pressure (see section 3). Evacuate all personnel from danger area. Do not approach area without self-contained breathing apparatus and protective clothing. Immediately cool cylinders with water spray from maximum distance, taking care not to extinguish flames. Remove sources of ignition if without risk. If flames are accidentally extinguished, explosive reignition may occur. Reduce corrosive vapors with water spray or fog. Stop flow of gas if without risk, while continuing cooling water spray. Remove all containers from area of fire if without risk. Allow fire to burn out. On-site fire brigades must comply with OSHA 29 CFR 1910.156.

(continued)

Figure 1-1 Praxair® Material Safety Data Sheet For Anhydrous Ammonia (cont.)

Product: Ammonia, Anhydrous   Form No.: P-4562-E   Date: October 1997

5. Fire Fighting Measures

(continued from previous page)

UNUSUAL FIRE AND EXPLOSION HAZARDS: Forms explosive mixtures with air and oxidizing agents. Heat of fire can build pressure in cylinder and cause it to rupture; no part of cylinder should be subjected to a temperature higher than 125 deg. F (52 deg. C). Ammonia cylinders are equipped with a pressure relief device. (Exceptions may exist where authorized by DOT—in this case, where the cylinders contain less than 165 pounds of product.) If leaking or spilled product catches fire, do not extinguish flames. Flammable and toxic vapors may spread from leak and could explode if reignited by sparks or flames. Explosive atmospheres may linger. Before entering area, especially a confined area, check atmosphere with an appropriate device. Reverse flow into cylinder may cause it to rupture. To protect persons from cylinder fragments and toxic fumes should a rupture occur, evacuate the area if the fire cannot be brought under immediate control.

HAZARDOUS COMBUSTION PRODUCTS: See section 10.

6. Accidental Release Measures

STEPS TO BE TAKEN IF MATERIAL IS RELEASED OR SPILLED:
DANGER! Toxic, corrosive, flammable liquefied gas under pressure (see section 3). Forms explosive mixtures with air (see section 5). Keep personnel away. Before entering area, especially a confined area, check atmosphere with an appropriate device. Use self-contained breathing apparatus and protective clothing where needed. Remove all sources of ignition if without risk. Reduce vapors with fog or fine water spray. Shut off flow if without risk. Ventilate area or move cylinder to a well-ventilated area. Contain spills in protected areas; prevent runoff from exposing personnel to liquid and vapors and contaminating the surrounding environment.

WASTE DISPOSAL METHOD: Prevent waste from contaminating the surrounding environment. Keep personnel away. Discard any product, residue, disposable container or liner in an environmentally acceptable manner, in full compliance with federal, state, and local regulations. If necessary, call your local supplier for assistance.

Figure 1-1 Praxair® Material Safety Data Sheet For Anhydrous Ammonia (cont.)

Product: Ammonia, Anhydrous   Form No.: P-4562-E .   Date: October 1997

7. Handling and Storage

PRECAUTIONS TO BE TAKEN IN STORAGE: Store and use with adequate ventilation. Firmly secure cylinders upright to keep them from falling or being knocked over. Screw valve protection cap firmly in place by hand. Store only where temperature will not exceed 125 deg. F (52 deg. C). Store full and empty cylinders separately. Use a first-in, first-out inventory system to prevent storing full cylinders for long periods.

PRECAUTIONS TO BE TAKEN IN HANDLING: Protect cylinders from damage. Use a suitable hand truck to move cylinders; do not drag, roll, slide, or drop. Never attempt to lift a cylinder by its cap; the cap is intended solely to protect the valve. Never insert an object (e.g., wrench, screwdriver, pry bar) into cap openings; doing so may damage the valve and cause a leak. Use an adjustable strap wrench to remove over-tight or rusted caps. Open valve slowly. If valve is hard to open, discontinue use and contact your supplier. For other precautions in using ammonia, anhydrous, see section 16.

For additional information on storage and handling, refer to Compressed Gas Association (CGA) pamphlet P-1, "Safe Handling of Compressed Gases in Containers," available from the CGA. Refer to section 16 for the address and phone number along with a list of other available publications.

8. Exposure Controls/Personal Protection

VENTILATION/ENGINEERING CONTROLS:
   LOCAL EXHAUST—Use a local exhaust ventilation system with sufficient airflow velocity to maintain concentration below the TLV in the worker's breathing zone.
   MECHANIC (general)—Not recommended as a primary ventilation system to control worker's exposure.
   SPECIAL—Use only in a closed system. An explosion-proof, corrosion-resistant, forced-draft fume hood is preferred.
   OTHER—None.
RESPIRATORY PROTECTION: Use air-supplied respirators for concentrations up to 10 times the applicable permissible exposure limit. For higher concentrations, a full-face, self-contained breathing apparatus operated in the pressure demand mode is required. Respiratory protection must conform to OSHA rules as specified in 29 CFR 1910.134.
PROTECTIVE GLOVES: Neoprene.
EYE PROTECTION: Wear safety glasses when handling cylinders; vapor-proof goggles and a face shield during cylinder change out or wherever contact with product is possible. Wear safety glasses when handling cylinders; vapor-proof goggles where needed. Select per OSHA 29 CFR 1910.133.
OTHER PROTECTIVE EQUIPMENT: Metatarsal shoes for cylinder handling. Protective clothing where needed. Select per OSHA 29 CFR 1910.132 and 1910.133. Regardless of protective equipment, never touch live electrical parts.

Figure 1-1 Praxair® Material Safety Data Sheet For Anhydrous Ammonia (cont.)

Product: Ammonia, Anhydrous   Form No.: P-4562-E   Date: October 1997

9. Physical and Chemical Properties

MOLECULAR WEIGHT: 17.031

EXPANSION RATIO: Not applicable

SPECIFIC GRAVITY (air=1): At 32 deg. F (0 deg. C) and 1 atm: 0.5970

SOLUBILITY IN WATER: Appreciable

GAS DENSITY: At 32 deg. F (0 deg. C) and 1 atm: 0.0481 lb/ft^3 (0.771 kg/m^3)

VAPOR PRESSURE: At 70 deg. F (21.1 deg C): 114.1 psig (786.7 kPa)

PERCENT VOLATILES BY VOLUME: 100

EVAPORATION RATE (Butyl Acetate = 1): High

BOILING POINT (1 atm): -28 deg. F (33.3 deg. C)

ACID/BASE pH : Not applicable

FREEZING POINT (1 atm): -107.9 deg F (-77.7 deg. C)

APPEARANCE, ODOR, AND STATE: Colorless gas at normal temperature and pressure; pungent, irritating odor.

10. Stability and Reactivity

STABILITY: Unstable: No   Stable: Yes

INCOMPATIBILITY (materials to avoid): Gold, silver, mercury, oxidizing agents, halogens, halogenated compounds, acids, copper, copper-zinc alloys (brass), aluminum, chlorates, zinc

HAZARDOUS DECOMPOSITION PRODUCTS: The normal products of combustion are nitrogen and water. Hydrogen may be formed at temperatures above 1,544 deg. F (840 deg. C).

HAZARDOUS POLYMERIZATION: May Occur: No   Will Not Occur: Yes

CONDITIONS TO AVOID: None known.

11. Toxicological Information

LC50 = 7338 ppm, 1 hr, rat.

12. Ecological Information

Ammonia, Anhydrous does not contain any Class I or Class II ozone-depleting chemicals. Ammonia, Anhydrous is not listed as a marine pollutant by DOT.

Figure 1-1 Praxair® Material Safety Data Sheet For Anhydrous Ammonia (cont.)

Product: Ammonia, Anhydrous   Form No.: P-4562-E   Date: October 1997

13. Disposal Considerations

WASTE DISPOSAL METHOD: Keep waste from contaminating surrounding environment. Keep personnel away. Do not dispose of unused quantities. Return cylinder to supplier.

14. Transport Information

DOT/IMO SHIPPING NAME: Ammonia, anhydrous, liquefied

HAZARD CLASS: 2.2 (domestic shipment), 2.3 (international shipment)

IDENTIFICATION NUMBER: UN 1005

PRODUCT RQ: 100 lbs (4.54 kg)

SHIPPING LABEL(s):   NONFLAMMABLE GAS (domestic);

TOXIC GAS, CORROSIVE (international)

PLACARD (when required): NONFLAMMABLE GAS (domestic);

TOXIC GAS, CORROSIVE (international)

SPECIAL SHIPPING INFORMATION: Cylinders should be transported in a secure position, in a well-ventilated vehicle. Cylinders transported in an enclosed, nonventilated compartment of a vehicle can present serious safety hazards.

Additional Marking Requirement: Inhalation Hazard (for international shipment)

Shipment of compressed gas cylinders that have been filled without the owner's consent is a violation of federal law [49 CFR 173.301 (b)].

15. Regulatory Information

The following selected regulatory requirements may apply to this product. Not all such requirements are identified. Users of this product are solely responsible for compliance with all applicable federal, state, and local regulations.

U.S. FEDERAL REGULATIONS:

EPA (Environmental Protection Agency)

CERCLA: Comprehensive Environmental Response, Compensation, and Liability Act of 1980 (40 CFR Parts 117 and 302):

Reportable Quantity (RQ): 100 lbs (45.4 kg)

SARA: Superfund Amendment and Reauthorization Act:

• **SECTIONS 302/304:** Require emergency planning based on Threshold Planning Quantity (TPQ) and release reporting based on Reportable Quantities (RQ) of extremely hazardous substances (40 CFR Part 355):

Figure 1-1 Praxair® Material Safety Data Sheet For Anhydrous Ammonia (cont.)

Product: Ammonia, Anhydrous   Form No.: P-4562-E   Date: October 1997

15. Regulatory Information (continued)

Threshold Planning Quantity (TPQ): 500 lbs (226.8 kg)
Extremely Hazardous Substances (40 CFR 355): None

- **SECTIONS 311/312:** Require submission of Material Safety Data Sheets (MSDSs) and chemical inventory reporting with identification of EPA hazard categories. The hazard categories for this product are as follows:

  IMMEDIATE: Yes   PRESSURE: Yes
  DELAYED: Yes   REACTIVITY: Yes
  FIRE: Yes

- **SECTION 313:** Requires submission of annual reports of release of toxic chemicals that appear in 40 CFR Part 372.
  Ammonia, Anhydrous requires reporting under Section 313.

40 CFR 68: Risk Management Program for Chemical Accidental Release Prevention: Requires development and implementation of risk management programs at facilities that manufacture, use, store, or otherwise handle regulated substances in quantities that exceed specified thresholds.

Ammonia, Anhydrous is listed as a regulated substance in quantities of 10,000 lbs (4536 kg) or greater.

TSCA: Toxic Substances Control Act: Ammonia, is listed on the TSCA inventory.

OSHA (OCCUPATIONAL SAFETY AND HEALTH ADMINISTRATION):

29 CFR 1910.119: Process Safety Management of Highly Hazardous Chemicals: Requires facilities to develop a process safety management program based on Threshold Quantities (TQ) of highly hazardous chemicals.

Ammonia, Anhydrous is listed in Appendix A as a highly hazardous chemical in quantities of 10,000 pounds (4536 kg) or greater.

STATE REGULATIONS:

CALIFORNIA: This product is not listed by California under the Safe Drinking Water Toxic Enforcement Act of 1986 (Proposition 65).

PENNSYLVANIA: This product is subject to the Pennsylvania Worker and Community Right-To-Know Act (35 P.S. Sections 7301-7320).

Figure 1-1 Praxair® Material Safety Data Sheet For Anhydrous Ammonia (cont.)

Product: Ammonia, Anhydrous   Form No.: P-4562-E   Date: October 1997

16. Other Information

Be sure to read and understand all labels and instructions supplied with all containers of this product.

OTHER HAZARDOUS CONDITIONS OF HANDLING, STORAGE, AND USE: *Toxic, corrosive, flammable, liquefied gas under pressure.* Do not breathe gas. Do not get vapor or liquid in eyes, on skin, or on clothing. (See section 3.) Have safety showers and eyewash fountains immediately available. Use piping and equipment adequately designed to withstand pressures to be encountered. Store and use with adequate ventilation at all times. *Prevent reverse flow.* Reverse flow into cylinder may cause rupture. Use a check valve or other protective device in any line or piping from the cylinder. *May form explosive mixtures with air.* Keep away from heat, sparks or open flame. Ground all equipment. Use only spark-proof tools and explosion-proof equipment. Store and use with adequate ventilation at all times. Use only in a closed system. Close valve after each use; keep closed even when empty. Keep away from oxidizing agents and from other flammables. *Never work on a pressurized system.* If there is a leak, close the cylinder valve. Blow the system down in an environmentally safe manner in compliance with all federal, state, and local laws, then repair the leak. *Never ground a compressed gas cylinder or allow it to become part of an electrical circuit.*

NOTE: *Prior to using any plastics, confirm their compatibility with ammonia, anhydrous.*

MIXTURES: When you mix two or more gases or liquefied gases, you can create additional, unexpected hazards. Obtain and evaluate the safety information for each component before you produce the mixture. Consult an industrial hygienist, or other trained person when you evaluate the end product. Remember, gases and liquids have properties that can cause serious injury or death.

HAZARD RATING SYSTEMS:

NFPA RATINGS:			HMIS RATINGS:	
HEALTH	=2 (gas)	3 (liquid)	HEALTH	=3
FLAMMABILITY	=1 (gas)	1 (liquid)	FLAMMABILITY	=1
REACTIVITY	=0 (gas)	0 (liquid)	REACTIVITY	=0
SPECIAL	None			

STANDARD VALVE CONNECTIONS FOR U.S. AND CANADA:
    THREADED: CGA-705, CGA-240 Standard,
                         CGA-660 Limited Standard
    PIN-INDEXED YOKE: Not applicable
    ULTRA-HIGH INTEGRITY CONNECTION: CGA-720

Use the proper CGA connections. **DO NOT USE ADAPTERS.** Additional limited-standard connections may apply. See CGA Pamphlet V-1 listed below.

Figure 1-1 Praxair® Material Safety Data Sheet For Anhydrous Ammonia (cont.)

Product: Ammonia, Anhydrous   Form No.: P-4562-E   Date: October 1997

16. Other Information (continued)

Ask your supplier about free Praxair safety literature as referenced on the label for this product; you may also obtain copies by calling 1-800-PRAXAIR. Further information about ammonia, anhydrous, can be found in the following pamphlets published by the Compressed Gas Association, Inc. (CGA), 1725 Jefferson Davis Highway, Arlington, VA 22202-4102, Telephone (703) 412-0900.

P-1 *Safe Handling of Compressed Gases in Containers*
V-1 *Compressed Gas Cylinder Valve Inlet and Outlet Connections*
Handbook of Compressed Gases, Third Edition

Praxair asks users of this product to study this Material Safety Data Sheet (MSDS) and become aware of product hazards and safety information. To promote safe use of this product, a user should (1) notify employees, agents and contractors of the information on this MSDS and of any other know product hazards and safety information, (2) furnish this information to each purchaser of the product, and (3) ask each purchaser to notify its employees and customers of the product hazards and safety information.

The opinions expressed herein are those of qualified experts within Praxair, Inc. We believe that the information contained herein is current as of the date of this Material Safety Data Sheet. Since the use of this information and the conditions of use of the product are not within the control of Praxair, Inc.,it is the user's obligation to determine the conditions of safest use of the product.

Praxair MSDSs are furnished on sale or delivery by Praxair or the independent distributors and suppliers who package and sell our products. To obtain current Praxair MSDSs for these products, contact your Praxair sales representative or local distributor or supplier. If you have questions regarding Praxair MSDSs, would like the form number and date of the latest MSDS, or would like the names of the Praxair suppliers in your area, phone or write the Praxair Call Center (**Phone:** 1-800-PRAXAIR; **Address:** Praxair Call Center, Praxair, Inc., PO Box 44, Tonawanda, NY 14150-7891).

Praxair is a trademark of Praxair Technology, Inc.

Praxair, Inc.
39 Old Ridgebury Road
Danbury, CT 06810-5113

Figure 1-1 Praxair® Material Safety Data Sheet For Anhydrous Ammonia (cont.)

MSDS Sections

Hazards Identification

Section 3 of Figure 1-1 identifies the hazards associated with anhydrous ammonia. The *Emergency Overview*, a feature of many more complete MSDS formats, provides the basic hazards and how to avoid them. This kind of summary of safety and health information needs to be included in written operating procedures and used for employee training. It rapidly communicates that anhydrous ammonia is harmful if inhaled and that rescue workers must wear self-contained breathing apparatus (SCBA).

Section 3 also lists threshold limit values in terms of a Threshold Limit Value—Time Weighted Average (**TLV-TWA**) and a Short Term Exposure Limit (**STEL**). A short-term exposure is usually a 15 minute exposure. The first value of 25 ppm (parts ammonia per million in air) is the time-weighted average exposure concentration for a conventional eight to ten-hour workday and a forty-hour workweek. As determined by the ACGIH, this exposure will have no long-term harmful effects for most workers handling anhydrous ammonia. The STEL value of 35 ppm is the 15 minute time-weighted average exposure that should not be exceeded at any time during a workday, even if the eight-hour TWA is within the permissible exposure limit of 25 ppm.

OSHA's terminology for TLV is **PEL** (Permissible Exposure Limit)—again, a time-weighted average. The federal OSHA PEL for anhydrous ammonia is 50 ppm. In this case, the manufacturer chose the more conservative value of 25 ppm, a common practice when providing exposure limits for toxic materials. Federal OSHA does not list a STEL or ceiling limit for ammonia. The reader is advised to determine whether his or her state OSHA office requires lower numbers than federal exposure limits. Note that some chemicals have OSHA "action levels" that are typically one-half the PEL. Action levels require special precautions that may or may not be spelled out on the MSDS. These and other exceptions to the general rules about exposure limits will cause many plant safety and health coordinators to seek professional guidance from an industrial hygienist or toxicologist.

Acute Overexposures

The hazards identification section also lists effects of a single or acute overexposure to liquid or gaseous ammonia via inhalation, skin contact, swallowing, and eye contact. The OSHA 200 log record keeper should remember that acute exposures are nearly always classified as injuries. We are told here that inhalation and skin/eye contact are the primary acute exposure hazards for ammonia.

Chronic Exposures and Other Effects

Chronic or repeated effects of ammonia overexposure are possible chemical pneumonia and kidney damage. Repeated exposure to ammonia may aggravate medical conditions like pulmonary disease and dermatitis.

Toxicological Studies

For ammonia, section 3 reports that no laboratory data exists linking ammonia with long-term effects on humans. Section 11 summarizes results of animal dose-response studies. The term, LC50 = 7338 ppm, 1 hr, rate means that the lethal (death causing) concentration (LC) for 50% of the rats studied was 7338 ppm for 1 hour.

Toxic Hazard Evaluation

The first question to raise is "Why is it important to estimate worker exposures to toxicants?" Among the reasons are the following:

- The facility needs to estimate or measure the level of employee exposure to potentially hazardous conditions and compounds in order to protect the employee's health.
- The exposure level must be compared with permissible exposure levels to determine compliance with regulations.
- Calculations and measurement is used to evaluate the effectiveness of engineering controls and personal protective equipment, such as ventilation.

There are several methods for evaluating toxic hazards in the workplace. Hazards can be evaluated through measurement or estimation (modeling). It is always best to determine exposures to toxicants through direct measurement. However, estimates must be used if direct measurement is impractical or impossible—for example, if the process equipment is not yet in place, or if release estimates of worst-case scenarios are needed. Estimates are also needed in locations in which measurements are difficult to take, such as in enclosed or confined spaces, above open containers, during drum and tank filling operations, and in places where spills may occur. When direct measurement of toxicants in the air *is* possible, some form of air sampling or monitoring is generally used.

Air Sampling and Monitoring

Two approaches to measuring toxic hazards are continuous monitoring and intermittent sampling. All air monitoring systems have limitations. Consultation services provided by professional industrial hygienists are highly recommended in order to effectively select and place air monitoring devices and equipment.

Sampling strategy includes consideration of a number of important questions, such as:

1. Do we take personal samples or area samples?
2. Who takes samples and/or where do we take samples?
3. When and with what frequency do we sample?
4. How many samples and how large are the samples needed to reliably measure the exposure? How long do we need to sample?

Note that personal samples provide the most accurate measure of a worker's exposure. For example, an organic vapor monitor placed directly in an employee's breathing zone will best measure the exposure while dispensing benzene. Some considerations for area samples are:

- Do we want to sample at the source of air contamination or a number of places in the room?
- If source sampling, do we want the worst possible exposure in a short time or a time-weighted average over the entire shift? How do we factor in evaluation of a control (air sampling when the air contaminant source is not active)?

Air monitoring products and analysis services are now available that simplify sampling of air contaminants. Figure 1-2 lists organic air contaminant compounds and the corresponding recommended sampling period for 3M air monitors.

Toxic Release Source and Dispersion Models

An important means of estimating the effects of toxic hazards is the use of toxic release source and dispersion models. There are generally two types of releases to the atmosphere: limited releases and massive (or catastrophic) releases. Limited releases are often caused by "leaks" in process equipment caused by cracks and/or holes in vessels, as well as leaky valves, pumps, and flanges. Leaks may result in acute or chronic employee expo-

3M 3500/3510 and 3520/3530 Compound List

†*Acetone (1.5)
*Acetonitrile (2)
Acrylonitrile (8)
Allyl Alcohol (8)
Amyl Acetate (8)
n-Amyl Alcohol (8)
s-Amyl Alcohol (8)
Benzene (8)
Benzyl Chloride (8)
Bromoform (8)
t-Butyl Acetate (8)
n-Butyl Acetate (8)
s-Butyl Acetate (8)
Butyl Acrylate (8)
n-Butyl Alcohol (8)
s-Butyl Alcohol (8)
t-Butyl Alcohol (8)
Butyl Cellosolve (8)
Butyl Glycidyl Ether (8)
p-tert Butyl Toluene (8)
Camphor (8)
Carbon Tetrachloride (8)
Cellosolve (8)
Cellosolve Acetate (8)
Chlorobenzene (8)
Chlorostyrene (8)
Chlorotoluene (8)
Chloroform (8)
Cumene (8)
Cyclohexane (6)
Cyclohexanol (8)
Cyclohexanone (8)
Cyclohexene (8)
n-Decane (8)
Diacetone-Alcohol (8)

o-Dichlorobenzene (8)
p-Dichlorobenzene (8)
Diisolbutyl Ketone (DIBK) (8)
p-Dioxane (8)
Enflurane (8)
Epichlorohydrin (8)
Ethyl Acetate (6)
Ethyl Benzene (8)
Ethylene Chlorohydrin (8)
Ethylene Dichloride (EDC) (8)
*Ethyl Ether (4)
Furfural (8)
Halothane (8)
n-Heptane (8)
n-Hexane (8)
iso-Amyl Acetate (8)
iso-Butyl Alcohol (8)
Isoflurane (Forane) (8)
Isophorone (8)
Isopropyl Acetate (8)
*Isopropyl Alcohol (4)
Mesityl Oxide (8)
Mesitylene (8)
Methyl Acrylate (8)
Methyl t-Butyl Ether
  (MTBE) (8)
Methyl Butyl Ketone
  (MBK) (8)
Methyl Cellosolve (8)
Methyl Cellosolve Acetate (8)
*Methylene Chloride
  (3530 only) (4)

†Methyl Ethyl Ketone (MEK) (8)
Methyl Isobutyl Ketone
  (MIBK) (8)
Methyl Methacrylate (8)
Methol Propyl Ketone (8)
Naphtha (VM&P) (8)
n-Octane (8)
Perchloroethylene (8)
Phenyl Ether (8)
n-Propyl Acetate (8)
n-Propyl Alcohol (7)
Propylene Dichloride (8)
Propylene Glycol Mono
  Methyl Ether (8)
Propylene Glycol Mono
  Methyl Ether Acetate (8)
Stoddard Solvent (8)
Styrene (8)
1,1,2,2,-Tetrachloroethane (8)
Tetrahydrofuran (8)
Toluene (8)
1,1,1-Trichloroethane (Methyl
  Chloroform) (6)
1,1,2-Trichloroethane (8)
Trichloroethylene (8)
*1,1,2-Trichloro-1,2,2-
  Trifluroethane (1)
†Vinyl Acetate (8)
Vinyl Toluene (8)
Xylene (8)
Total Hydrocarbons as n-Hexane

The number in parenthesis is the recommended sampling period in hours. This time has been estimated using the capacity of the 3510 organic vapor monitor, a relative humidity of <70% and the 1995 ACGIH TLVs. Use of the 3530 allows the sampling time to increase.

Because of their high vapor pressures (low Boiling Points), the starred (*) compounds are best sampled initially with the 3520 or 3530 monitor (with back-up section). Subsequent sampling may be done with the 3500/3510 monitors if determined, by 3520 results.

†NOTE: certain compounds (e.g. acetone, methyl ethyl ketone, vinyl acetate, etc.) may show a decreased recovery when sampled in high relative humidity. Refrigerate and/or expedite for analysis to help ensure accurate results.

Figure 1-2 3M™ Air Monitoring Systems[3]

sure. Massive releases are often caused by explosions or other causes of total equipment failure.

Examples of release sources are open rupture disks, significant holes in vessels, cracked or broken pipes, etc. In order to predict the impact once a chemical is released, a source model is used to make a reasonable estimate of the rate at which material is released. Once the release rate is determined, dispersion models are used to predict the area affected by the released toxic "plume" or "puff" and can estimate the concentration of gas/vapor at various points throughout the chemical cloud.

Use of release and dispersion models are important for several reasons. First and foremost, these models provide the possible impact of a significant release on the facility, employees, and the community. Thus, they aid the facility in developing a release prevention plan using the elements of OSHA's Process Safety Management regulation. Release and dispersion models also help the facility in emergency planning and response. Note that EPA requires plants with regulated materials to coordinate their emergency response program with the corresponding county or local community plan (see Chapter 8). Cooperation in planning such matters as training, alerting local public emergency responders, documenting proper first aid and medical treatment, and use of emergency equipment is essential to an integrated response to a toxic release or a fire/explosion.

EPA Risk Management Program (RMP)

The EPA Risk Management Program (RMP) regulation requires modeling in order to estimate worst-case and more likely offsite consequences. EPA provides an offsite consequence guidance document and a software model (RMP*Comp) that can be used to determine the distance to an offsite impact. Appendix A is a step-by-step guide to using the EPA RMP Guidance document (available from the EPA website[4]) in order to estimate worst-case and alternative release scenarios.

Using Computerized Release and Dispersion Models

Several software packages are available for the safety manager to model toxic releases. Some use proprietary models to calculate the effects of a release; others can be downloaded for free. Appendix B provides a list of descriptions and a comparison of the following systems:

*EPA's RMP*Comp*
ALOHA®
SAFER SYSTEMS®

Release source and dispersion models provide tools for evaluating the effects of a toxic release, and thus information to design appropriate controls—prevention and mitigation measures. By identifying and quantifying the potential "footprint" for a release, facility and corporate management can better determine the risks involved and set priorities for investments in toxic hazard controls.

Toxic Hazard Controls

Once real or potential health hazards to employees have been identified and evaluated, functional exposure control measures must be developed and installed.

Any discussion of hazard or exposure control must begin with a few questions.

- What are we trying to control?
- What are we trying to protect?
- What level of protection is adequate?
- What method should be used?
- When (in what situations) should the method be considered?

In discovering what a control measure is, a fundamental concept and principle that is of paramount importance must be introduced: the difference between a chemical's toxicity and a chemical hazard.[5] No control measure can reduce the *toxicity* of a particular substance. Proper control measures can, however, reduce the *hazard* a particular substance presents to the user.

Chemical Hazard Control

Safety Precedence Sequence

Whether controlling chemical or physical hazards, it is important to do the most effective job possible *removing* the hazard. There is a general priority to corrective action called the *safety precedence sequence*:

Safety Precedence Sequence
(In order of decreasing effectiveness)

1. Use engineering controls—design the solution to reduce or eliminate the hazard.
2. Install automatic safety devices like pressure relief valves, rupture disks, and fail-safe process control interlocks.

3. Install automatic safety warnings—alarms, sirens, console lights, permanent signs, etc.

4. Develop or improve written procedures—including use of specified personal protective equipment (PPE).

5. Take some personnel actions—for example, increase supervision, classroom, or on-the-job training; adjust working hours; and take disciplinary action.

6. Identify residual risks to the appropriate level of management—if the hazard cannot be removed, measures must be taken to reduce exposure of the worker to the hazard.

The Safety Precedence Sequence (SPS) is useful for eliminating or controlling hazards—especially when it is used up front for design purposes or for corrective action following an accident investigation. Note that the strategies are listed in order of decreasing effectiveness. For example, use of ventilation is the most effective control method for potential inhaled air contaminants because it generally "engineers-out" the hazard. Self-contained breathing apparatus (SCBA), respirators, hoods, and other personal protection equipment (PPE) should be considered as a last line of defense against toxicants. Unfortunately, PPE is used more frequently and often automatically because this method is cheaper in the short-term. All too often the worker ultimately pays the price of inhaling these substances due to lack of, improper, or ill fitting PPE, or PPE that is used beyond its effective life.

SPS needs to be used in a creative and careful way. If the hazard cannot be removed by engineering controls, measures must be taken to reduce or otherwise control exposure of the employee to the hazard.

Chemical Hazard Control Options

It is useful to list hazard control options and techniques in an order that roughly parallels the intent of the safety precedence sequence:[6]

Approach	Example
1. Substitution, Intensification, and Attenuation	Use chemicals or solvents that are less toxic. Reduce inventories and quantities in the process. Change from large batch reactors to a continuous process. Reduce temperatures and pressures. Reduce boiling point using vacuum. Refrigerate storage vessels.

2.	Engineering Design Controls	Contain toxic chemicals in closed systems. Design process control features to include emergency shutdown. Local exhaust and area ventilation. Dikes, buildings and other equipment and personnel enclosures. Automatic quench systems for exothermic reactions. Operate at process conditions where runaway reaction is not possible.
3.	Countermeasures and Emergency Response	Automatic or remotely actuated water sprays, curtains, foams, etc. Emergency shutdown equipment and procedures. Evacuation and shelter-in-place procedures. Unit and site communication systems.
4.	Early Detection and Warning	Personal and area toxic vapor sensors/monitors. Detection and alarm systems.
5.	Policies and Procedures	Written operating and maintenance procedures like Job Safety Analysis. Site-specific hazard communication and process safety training. Equipment inspection and test programs (mechanical integrity). Management of change systems. Audits and inspections. Provisions for site security.
6.	PPE and Good Housekeeping	Face masks, chemical goggles, and safety glasses with side shields, hearing protection, aprons, and chemical impervious gloves and rated chemical suits. Contain all toxic liquids, vapors and dusts. Provisions for water, steam and other clean-in-place systems. Lines for flushing and cleaning. Sewer systems with provisions for emergency containment.

The best place to prevent a release is at the source. There are two basic approaches to mitigation:

1. Prevention—reducing the likelihood of a catastrophic event through use of engineering controls to prevent a release
2. Protection—reducing the magnitude of the release and/or the exposure of employees, neighbors, and properties

Substitution, Intensification and Attenuation

Examine opportunities to replace hazardous chemicals with those having lower toxic and polluting properties wherever possible. Consider substitution for chemicals such as chlorinated and chlorofluoro-carbons and fluorinated, nitrogen, or sulfur-containing organics.[7]

Process intensification and attenuation are terms meaning special kinds of engineering control. Process intensification has to do with minimizing or reducing the quantity of chemical in a process, thus reducing the amount of toxic material or energy that might be released during an incident. The term may also be used to describe application of new technologies that not only reduce the quantity of chemicals used but also may eliminate entire pieces of equipment (like large batch reactors) or steps in the process. Examine unit operations that involve reactors, distillation units, heat exchangers, plus gas/liquid and other separation devices for opportunities to apply process intensification technology.

Attenuation means using chemicals under conditions which make them less hazardous. For example, process designers will apply vacuum to reduce the boiling point of a chemical mixture. Also consider reflux cooling (e.g. condensing an excess of solvent to control the temperature of a chemical reaction).

Engineering Design Controls and Other Mitigation Measures

A fundamental principal in chemical process design is to contain all materials in closed systems. Process controls should include quench or other rapid emergency shutdown features.

Enclosures are often either overlooked as a way to control chemical hazards or the enclosure is located too close to the source of the release. This is especially true of older process buildings where the control room is often located in the middle of an array of pipes, pumps, and reactors. In any case the control room should be placed under higher pressure relative to the process area to prevent unwanted gases and vapors from entering the space.

Local and dilution ventilation is the most common way to protect workers from toxic vapors. Hoods and local exhausts (commonly called "elephant trunks") are common ways to contain and exhaust hazardous substances. In addition to health protection, dilution ventilation also may provide fire protection and worker comfort. In general, dilution ventilation is not as satisfactory for health hazard control as is local exhaust ventilation. However, there are situations in which dilution ventilation is the easy choice, such as when local exhaust is prohibited by the process or is prohibitively expense. The heating, ventilation and air conditioning (HVAC) person should be careful of basing economic decisions solely on initial equipment costs—for example, the initial cost of a local exhaust system. Dilution systems usually exhaust large volumes of heat from a building. With the current high energy costs, the economics of dilution ventilation have changed considerably. Note also that simple fixes such as increasing the re-circulation rate relative to outside air can backfire and cause a whole host of worker health problems.

There are four other limiting factors when it comes to diluting hazardous gases and vapors:

1. The quantity of contaminant must not be too large.
2. There needs to be sufficient distance between the hazard source and the worker.
3. The chemical should have relatively low toxicity.
4. The flow of contaminant should be relatively uniform over time (as opposed to flow that varies widely or is intermittently on-off).

Note that general ventilation is generally not effective for control of hazardous dusts.

The release source model provides a tool for designing appropriate release prevention and mitigation measures, such as dikes, enclosures, berms, drains, sumps (all passive systems) and active systems, such as sprinkler systems, deluge systems, water curtain, neutralization, excess flow valves, flares, scrubbers and, emergency shutdown systems.

Dispersion models predict the area affected by a toxic plume or puff and can determine the concentration of gas/vapor at various points throughout the release cloud.

It is a good idea to develop a new toxic release and dispersion model after prevention and protection changes have been made. The modeling procedure is continued until hazards are reduced to an acceptable level.

Hazardous Chemical Examples

In this section, we will use the following specific chemical examples to illustrate the importance of knowing what to look for before beginning measurement and prevention activities. Research for application of the chemicals by industry comes from the Environmental Defense Scorecard.[8] Reversible and irreversible effect information has been summarized by consulting the NIOSH/OSHA Pocket Guide to Chemical Hazards,[9] the Guide to Occupational Exposure Values,[10] and the material safety data sheets for the specific chemicals. The toxic effect information is divided into three columns: TWA and STEL values, effects that may or may not be reversible, and irreversible effects. TWA and STEL values are either ACGIH or OSHA numbers (whichever is the most restrictive).

Why these ten chemicals? They are all high volume chemicals (HVCs) with production exceeding one million pounds annually in the U.S. All are covered by OSHA, EPA, and Department of Transportation (DOT) regulations affecting material storage, handling, and shipment. Your employees are the first line of chemical exposure. Your neighbors, including other nearby commercial and industrial operations, local communities, emergency responders, and public officials are other "stakeholders" concerned with possible chemical pollution and exposure. Organizations such as Environmental Defense communicate emissions/release data for these and other materials that companies are required to report to the EPA. TRI (Toxics Release Inventory) data and other technical information are often used to sensitize the public about the health effects and risks of chemical exposure. So this list is representative of chemicals that manufacturing and processing facilities should be concerned about.

Table 1-1 Toxic Effects On the Body for Ten Chemicals

Chemical, OSHA Standard	ACGIH/OSHA[1]		May/may not be reversible					Irreversible		
	TWA[2]	STEL[3],	R	S/E	L,K	B	CNS	C	G	R,BD
Ammonia, anhydrous (1910.111)	25	35	X	X	K?	—	—	—	—	—
Benzene (1910.1028)	0.5	2.5	X	X	X (reversible)	X	X	X	X	X
Chlorine	0.5	1	X	X	—	—	—	—	—	—
Ethanol, denatured	1000	—	X	X	X	—	X	—	—	X
Ethylene oxide (1910.1047)	1	5	X	X	—	X	X	X	X	X
Hydrogen chloride	—	5	X	X	—	—	X	—	—	—
Hydrogen sulfide	10	15	X	X	—	—	X	—	—	—
Toluene-2,4-di-isocyanate	0.005	0.02	X	X	—	—	—	X	—	—
Vinyl acetate	10	15	X	X	—	—	X	?	—	—
Vinyl chloride (1910.1017)	1	5	X	X	X	X	X	X	X	X

Effects that may or may not be reversible
R   Respiratory system (pulmonotoxic), lung effects
S/E   Skin/Eye (dermatotoxic) effects
L,K   Liver (hepatotoxic) and/or Kidney (nephrotoxic) effects
B   Blood (hemotoxic) effects
CNS   Central Nervous System (neurotoxic) effects

Effects that are irreversible
C   potential Cancer causing (carcinogen)
G   potential Genetic effects (mutagen), chromosome damage
R, BD potential - Reproductive system effects, Birth Defects

(1) American Conference of Governmental Industrial Hygienists, Inc. (ACGIH ®) Threshold Limit Value (TLV) or OSHA Permissable Exposure Limit (PEL), which- ever is more restrictive.
(2) **TWA:** Time-weighted average exposure concentration, parts per million in air, for a conventional 8-hour or up to a 10-hour workday and a 40-hour work- week.
(3) **STEL:** Short term exposure limit, parts per million, usually a 15-minute time- weighted average exposure that should not be exceeded at any time during a workday, even if the 8-hour TWA is within permissible exposure limits.

Anhydrous ammonia—Chemical Abstracts Service (CAS) Number 7664-41-7

Anhydrous ammonia is a colorless gas at normal temperatures and pressures with a penetrating pungent odor. Ammonia is broadly used in industry, finding application in chemical laboratories, circuit board manufacture, pharmaceuticals, pulp and paper processing, and the manufacturing of rubber.

Ammonia affects the respiratory system, skin and eyes as an irritant. Chronic or repeated effects of ammonia overexposure include possible kidney damage.[11] These effects may or may not be reversible. Anhydrous ammonia is usually not acutely lethal. It has not been positively linked to cancer, genetic changes, reproductive system effects, or birth defects.

Benzene—CAS Number 71-43-2

Benzene is a colorless liquid with an aromatic odor. In spite of the hazards involved, benzene continues to be used as a solvent in the manufacture of herbicides and pesticides, pharmaceuticals, as well as wood stains and varnishes.

Benzene has all of the reversible and irreversible effects listed. Acute inhalation overexposure to benzene will initially act as a narcotic, possibly leading to coma in extreme cases.[12] If involved in a fire, benzene will decompose to produce toxic gases. Aspiration into the lungs, following ingestion, can result in severe damage to the lungs; death may result.

Chronic exposure to benzene causes serious health effects by all routes of exposure. Frequent oral and inhalation exposure causes severe effects on the blood system, including damage to the bone marrow, leading to a decrease in production or changes to red and white blood cells. Benzene is a confirmed human carcinogen. Human mutation data have been reported for benzene. Symptoms of chronic exposure by most routes may be delayed for months to years after exposure has stopped.

Chlorine—CAS Number 7782-50-5

Chlorine is an amber liquid or greenish-yellow gas with a characteristically irritating odor. Chlorine is another workhorse chemical having broad application. Industries using this chemical include semiconductor and printed circuit board manufacturing, pulp and paper manufacturing, chemically inert liquids, and water treatment.

Exposure to chlorine affects the respiratory system, eyes, and skin as an irritant.[13] Acute inhalation of high concentrations can result in unconsciousness and death. Chronic or repeated chlorine overexposures by inhalation can result in emphysema and erosion of tooth enamel.

Genetic effect data have been reported during experimental studies on tissue using relatively high exposure levels. Chlorine has not been reported to cause mutagen effects in humans and it is not considered to be carcinogenic.

Ethanol (Ethyl Alcohol)—CAS Number 64-17-5

Ethanol is a colorless liquid with a slightly fruity odor. Ethanol has many solvent applications including industries like electroplating, paint manufacture, pharmaceuticals, printing, wood stains and varnishes, and the production of antifreeze agents. In addition, ethanol, distilled from fermented corn, is gaining popularity as a gasoline additive/replacement.

Ethanol has by far the highest TLV for all of the chemicals listed. An emergency overview warns of eye damage and skin irritation, central nervous system depression, and kidney/liver damage as primary health hazards.[14] Ethanol may cause irritation of the digestive tract and systemic toxicity with acidosis. Advanced stages of ethyl alcohol consumption may cause collapse, unconsciousness, coma and possible death due to respiratory failure. Reversible and irreversible effects are possible. Chronic exposures may cause dermatitis, reproductive, and birth defects.

Ethylene Oxide—CAS Number 75-21-8

Ethylene oxide is a colorless liquid or gas with an ether-like odor, irritating at high concentrations. It is commonly used in the production of pesticides and by hospitals for sterilizing.

Ethylene oxide has many potential hazards. It irritates the eyes and respiratory tract. Acute exposures if inhaled may be fatal or cause lung injury and the delayed onset of pulmonary edema. Depending on the degree of exposure, there may be stinging of the nose and throat, coughing, chest tightness, nausea, vomiting diarrhea, light headed feeling, weakness, drowsiness, cyanosis, loss of coordination, convulsions, and coma. A dilute solution of ethylene oxide may penetrate the skin, producing a chemical burn.

Repeated exposure to ethylene oxide may result in contact dermatitis to exposed workers. Reports of high recurrent exposures to ethylene oxide vapor have indicated toxicity to the central nervous system. Toxicological information indicates cancer causing, genetic changes, reproductive and birth defects. In humans, an increased occurrence of leukemia and stomach cancer has been reported by a group of Swedish investigators.[15]

Hydrogen chloride—CAS Number 7647-01-0

Hydrogen chloride or hydrochloric acid is a colorless gas with an irritating pungent odor. It is broadly used in a variety of industries including chemical laboratories, circuit board and semi-conductor manufacturing, electroplating, integrated iron and steel manufacturing, and the production of pH regulating agents.

Potentially reversible effects of hydrogen chloride include respiratory system, skin, and eye irritation/damage. Breathing of vapor or mist may be fatal. Symptoms may include severe irritation, and burns to the nose, throat, and respiratory tract. Central nervous system excitation and later depression is a sign of exposure to this material through breathing, swallowing, and/or passage of the material through the skin.[16] Irreversible effects have not been reported.

Hydrogen sulfide—CAS Number 7783-06-4

Hydrogen sulfide is a colorless gas with a strong odor (low odor at high concentrations) of rotten eggs; a liquid at high pressures and low temperatures. This chemical is used to produce sulfide ore activators and floatation agents, digesting chemicals for pulp and paper processing, and as a reducing agent for nonmetallic inorganic materials.

Like hydrogen chloride, hydrogen sulfide health effects are all potentially reversible. Acute exposures via inhalation are harmful to the respiratory system and may be fatal.[17] Symptoms and effects include headache, dizziness, vertigo, giddiness, confusion, chest pains, olfactory fatigue, unconsciousness, and even death. Hydrogen sulfide depresses activity of the central nervous system, which can cause respiratory paralysis as well as brain stem and cortical damage. Chronic exposures may cause nausea, vomiting, weight loss, persistent low blood pressure, and loss of the sense of smell.

Toluene-2, 4-di-isocyanate (TDI)—CAS Number 584-84-9

Toluene-2, 4-diisocyanate (TDI) is a colorless, yellow, or dark liquid or solid with a sweet, fruity, pungent odor. TDI is often found in adhesives and coatings, sealers, elastomers, as well as flexible and rigid foams.

The information for TDI includes other forms of the chemical (mixed isomers). Vapor pressure of the liquid is very low (0.025 mm Hg @ 25 deg. C), however the exposure limits are the lowest of the ten chemicals listed. Results of exposure include respiratory and eye-skin irritation/ damage and long-term, irreversible chemical sensitivity.[18] Acute TDI vapor exposure by inhalation may cause irritation of the mucous membranes of the nose, throat, or trachea; breathlessness; chest discomfort; difficult breathing; and reduced lung function. Pulmonary edema (fluid in lungs) may occur and sensitization may lead to asthma-type spasms of the bronchial tubes and difficulty in breathing. Chronic overexposure effects may also cause lung damage and permanent sensitization in some individuals. Cross-sensitization to different isocyanates may occur. Exposure through the skin may lead to pulmonary sensitization. TDI is listed by ACIGH as possibly cancer-causing to humans based on limited evidence of carcinogenic effects on rats.

Vinyl acetate—CAS Number 108-05-4

Vinyl acetate is a colorless liquid with a sweet fruity smell, irritating to some individuals. Vinyl acetate is an intermediate in the manufacture of acrylic fibers. It is also found in many adhesives, rubber, impregnation agents, as well as in pulp and paper manufacturing.

Problems associated with vinyl acetate exposure are mainly concerned with respiratory and skin/eye irritation. Levels above 20 ppm will likely cause nose and throat irritation resulting in coughing and hoarseness.[19] High concentrations can cause drowsiness and lack of coordination, a central nervous system effect. Inhalation overexposure at 600 ppm in air using rats and mice produced irritation and carcinomas in the nose and airway of some animals. There is insufficient data to classify vinyl acetate as cancer-causing to humans.

Vinyl chloride—CAS Number 75-01-4

Vinyl chloride is a sweet smelling, colorless gas. The monomer is a starting material for the PVC (poly vinyl chloride) industry and also is used to impregnate paper and other fibers.

From Table 1-1, we see that vinyl chloride is potentially harmful across the entire table. Exposure to high vapor concentrations may produce central nervous system effects, such as dizziness, headache, nausea, dulled vision, and drowsiness.[20] Even higher concentrations may lead to lung irritation, heart beat irregularities, liver and kidney damage, decreased blood clotting, unconsciousness, and even death. Vinyl chloride is a confirmed human carcinogen. Chronic effects of overexposure may lead to liver and kidney damage and cancer of the liver, brain, nervous system, lung, bile tract, and blood system. In addition, decreased ability of the blood to clot has been reported in humans. Decreased sex drive in men, and irregular menstrual periods in women, have been reported in persons exposed over a long period of time to vinyl chloride. At levels that caused adverse effects in the mother, reproduction effects that include miscarriage, decreased birth weight, and delayed skeletal development of the offspring have been reported.

Implications for Chemical Safety Management

When dealing with toxic chemicals, it is important to know what you are looking for and the precise nature of the hazard. Do not prescribe PPE as a quick fix before estimating the dose level and duration of actual or potential exposure to harmful chemicals. Take advantage of air sampling and monitoring systems that will help quantify the hazard.

Request a current version of the MSDS from your supplier. Using the CAS number, check other MSDS sources such as the Vermont SRI.[21] From the MSDS, study how specific chemicals enter the body. Draft a summary of the harmful effects to be used during employee orientation and hazard communication training. It is a good idea to purchase a toxicology reference book such as Sax's Dangerous Properties of Industrial Materials.[22] If using an industrial hygiene consultant, involve key employees and participate in the monitoring and analysis as much as possible.

Apply the safety precedence sequence and emphasize engineering controls to eliminate the hazard. Seek to control emissions and personnel exposures at the source as opposed to relying on, for example, thermal oxidizers and large chemical sewers.

Finally, get the most out of air dispersion models to estimate the potential offsite impact of a chemical release (Appendix A and B). Avoid technical jargon when explaining the hazards and prevention/mitigation activities to employees and the public. Listen to your audience before addressing the issues to reduce the emotion involved.

References

1. Code of Federal Regulations, 29 CFR1910.1200 Hazard Communication, (d), U.S. Government Printing Office, Washington, D.C. 20402.

2. Reprinted with permission, Praxair Technology, Inc. Copyright 1979, 1985, 1989, 1992-1993.

3. Technical Data Bulletin 1028, Organic Vapor Monitor Sampling and Analysis Guide, 3M Occupational Health and Environmental Safety, St. Paul, Minnesota, http://www.mmm.com/occsafety.

4. www.epa.gov/ceppo

5. C.L. Ley, M.D., 3M Industrial Hygiene Mini-seminar, 1992.

6. Adapted from Table 3-8, p. 69 and Table 5-4, p.150, Daniel A. Crowl and Joseph F. Louvar, *Chemical Process Safety: Fundamentals with Applications*, Prentice Hall International Series in the Physical and Chemical Engineering Sciences, Prentice-Hall, Inc., Division of Simon & Schuster, Englewood Cliffs, New Jersey 07632, 1990.

7. Adolf W. Gessner, "Switch Materials to Prevent Pollution," *Chemical Engineering Progress*, December 1998.

8. Chemical Profiles, Environmental Defense Scorecard, www.edf.org.

9. U.S. Department of Health and Human Services, Public Health Service, Centers for Disease Control, National Institute for Occupational Safety and Health (publication No. 78-210) and the U.S. Department of Labor, Occupational Safety and Health Administration, 1978.

10. American Conference of Governmental Industrial Hygienists, Inc., ISBN: 1-882417-41-0, 2001.

11. Praxair MSDS section 2.

12. MSDS, 1998, AIRGAS INC., 259 Radnor-Chester Road, Suite 100, Radnor, PA 19087.

13. Chlorine MSDS, 1998, AIRGAS INC.

14. Ethanol MSDS, 1998, Acros Organics N.V., One Reagent Lane, Fairlawn, NJ 07410.

15. Ethylene Oxide MSDS, 1998, PRAXAIR, INC.

16. Hydrogen Chloride Anhydrous MSDS, 1998, Ashland Chemical Co, P.O. Box 2219, Columbus, OH 13216.

17. Hydrogen Sulfide MSDS, 1998, PRAXAIR, INC.

18. Toluene Diisocyanate (mixed isomers) MSDS, 1999, BASF Corporation, 1609 Biddle Avenue, Wyandotte, Michigan 41892.

19. Vinyl Acetate Monomer MSDS, 1992, Phillips 66 Company, Bartlesville, Oklahoma, 74004.

20. Vinyl chloride monomer MSDS, 1992, Phillips 66.

21. Vermont Safety Information Resources, Inc., http://hazard.com

22. Richard J. Lewis, Sr., *Sax's Dangerous Properties of Industrial Materials*, 10th edition, American Conference of Governmental Industrial Hygienists, Inc. publication #0680.

Fires and Explosions

Fires

Fires and explosions are the most damaging chemical plant accidents throughout the U.S. and perhaps worldwide. An effective fire prevention plan is needed by every facility. The protection of lives and the continuation of business are the primary goals of such a plan.

First, property loss and business interruption costs represent a huge business expense. An equally large amount of money is spent annually on fire prevention and emergency planning/emergency response to fires and explosions. If fires and explosions are such an international "epidemic," and prevention money is well spent, why hasn't this tremendous source of tragedy been stamped out? A study by commercial and industrial property insurer FM Global of major causes of fires and explosions revealed that potential ignition sources are everywhere in most facilities and the majority of all loss can be prevented. (See Table 2-1.)

Table 2-1 Top 10 Ignition Sources at Properies Insured by FM Global (1989 to 1998)[1]

Cause	Number	Gross Loss U.S. Dollars
Electricity	1,497	$787,354,883
Hot Work	565	$696,302,342
Hot Surfaces	597	$482,883,397
Arson	759	$362,930,132
Spontaneous Ignition	313	$285,533,985
Exposure	232	$225,647,028
Friction	289	$223,008,409
Burner Flame	238	$176,255,610
Overheating	487	$140,149,881
Smoking	280	$122,107,182

© 2000 Factory Mutual Insurance Company
Reprinted with permission from "Ignition Sources: Recognizing the Causes of Fire."

The second reason fire prevention plans are essential is that fires and explosions can cause significant environmental problems. Whether the scenario is large or small, problems such as the following can result from these accidents:

- Production of toxic gas and dust from combustion and decomposition products
- Thermal plumes that carry toxic materials long distances
- Ground water and disposal problems caused by large quantities of contaminated water used in extinguishing the fire

What follows in this chapter examines the hazards of flammables and combustibles and the prevention of fires and explosions. This information pertains to more facilities than chemical plants—many industries and organizations use large quantities of flammable and combustible liquids. These materials have many uses, including the following:

- Dip tanks
- Cleaning metal parts
- Coating operations and drying ovens
- Spray booths
- Paint thinners
- Solvents in chemical mixing, milling, and other processing

Basic Fire Theory

The classic fire triangle used in fire training has been replaced. There are actually four components of a fire: fuel, oxygen, heat, and a chemical chain reaction. These four sides made up the fire tetrahedron and interact in the combustion process.

During combustion, heat energy is produced by a chain reaction of liquid phase fuel, a gas phase fuel, or a combination of both. The fuel is rapidly oxidized by the oxygen in the air surrounding the flame. Think of the gradual rusting of metal as an example of slow oxidation. Thus a fire is actually a combustion reaction intense enough to generate heat and light.

Theory experts classify fires into two broad forms: flame and surface fire. Flame fires may be one of two types[2]:
- "Premixed flame fires exist in a gas burner or stove and are relatively controlled, operating at high temperature.

- "Diffusion flame fires refer to gases burning on mixed vapors and air. Controlling these fires is difficult."

Try to light a stick of wood (combustible fuel), 1 inch thick, with a match. It doesn't work because there is insufficient heat to produce an ignitable vapor just above the wood surface. Thus surface fires actually burn in the vapor phase above a solid fuel. The danger in surface fires is the frequent presence of deep-seated embers that provide combustion heat and may re-ignite the flames if not fully extinguished.

Controlling and Extinguishing Fires

Fire control involves removing one or more of the sides of the fire tetrahedron:

1. Remove the heat more rapidly than the fire generates it. From a practical standpoint, this means the extinguishing agent (e.g. water) must be directly applied to the burning fuel.

2. Exclude air, choking off the oxygen supply. Examples here are rolling on the ground to extinguish burning clothing or dumping sand on a dying campfire.

   Note that "smothering a fire can be ineffective if the substance burning contains its own oxygen supply, such as found in ammonium nitrate or nitrocellulose. In addition, smothering does not work where deep-seated embers exist in materials such as wood, rags, and large rolls or skids of paper."[3]

   Carbon dioxide or CO_2 extinguishers work on this principle of displacing air. They work well for gas, grease, or solvent fires that are contained in a non-combustible enclosure such as over a gas stove burner or a laboratory vapor hood. CO_2 extinguishers are not effective for extinguishing combustible materials because of the difficulty in reaching deep-seated embers.

   Light Water[TM4] and other fire-fighting foams also work on the principle of excluding air from a fire. For example, these extinguishing agents float on top of the burning surface of flammable liquid fuels and cut off the oxygen supply needed to sustain the fire.

3. Fuel can be removed such that there is not enough heat for ignition, but once started, removing fuel from a fire can be a difficult and dangerous proposition.

   A safe and effective way to remove liquid or gaseous fuel like propane burning from a pipe is simply to shut off the supply.

Another method of controlling the fuel supply to a fire involves adding an excess of air to a mixture of fuel gases or vapors. This dilutes the fuel supply below the lower flammable limit (LFL). This ventilation method is normally applied to fire prevention and not as a means of extinguishing a fire.

4. Stop or slow the chain reaction. This action slows the generation of free radicals (necessary for a flame fire). "Extinguishing agents like dry chemicals (sodium or potassium bicarbonate, ammonium phosphate, and others) and halogenated hydrocarbons (chlorine, bromine, or fluorine) remove the free radicals in these branched-chain reactions from their normal function as a chain carrier." [5]

Classification of Fires and Extinguishers

The National Fire Protection Agency (NFPA) sets forth four general classifications of fires. These classes define the types of combustible fuels involved and the extinguishing agent needed to fight the fire.

Class A Fires

Ordinary combustible fuels like wood, paper, sawdust, rags, and trash are the primary source of class A fires. Think of "A" as standing for the acronym "ashes" (hidden, glowing embers which can re-ignite), the primary hazard involved with extinguishing these fires. Thus, the smothering and cooling effects of water are most effective in putting out class A fires. So-called ABC or all-purpose extinguishers contain a dry chemical, which provides rapid knockdown of the flames and formation of a coating to prevent oxygen from re-igniting the fuel. Follow up with water.

Class B Fires

Class B fires involve burning gas or vapor-air mixture over the surface of flammable or combustible liquids. Examples of class B fuels include propane, gasoline, kerosene, oil base paints, and many organic solvents like ethyl alcohol and methyl ethyl ketone (MEK). Here the "B" stands for the hazard "boiling," because it is the rapid boiling of the liquid that generates the burning vapor-phase fuel to produce heat and even more vapor. Limiting air (oxygen) and/or the chain reaction effects will control the fire. Streams of water may only spread the fire, since gasoline and many organic solvents float on water. Water fog nozzles are helpful in removing heat and thus controlling class B fires, but will not totally extinguish the flames. Extinguishing agents like regular or multi-purpose dry chemical, CO_2, foam, or halogenated agents are recommended.

Class C Fires

Class C fires occur in or near energized electrical equipment. The acronym associated with "c" is "current" because live electrical current is the most hazardous aspect of this type of fire. Foam or water should not be used because they are good conductors and may facilitate an electrical shock. De-energizing the circuit usually reduces this type of fire to class A, such as burning combustibles like insulation from the cord. Dry chemicals, CO_2, and halogenated agents are used for class C fires.

Class D Fires

This type is a special class of fire involving combustible metals like magnesium, titanium, zirconium, lithium, potassium, and sodium. Here, there is no acronym for "D," but the hazard is a potential chemical reaction between the burning metal and extinguishing agent. This hazard may well accelerate the intensity of the fire.

Other Fires

For other fires that involve certain combustible metals or reactive chemicals, consult NFPA 49 *Hazardous Chemicals Data,* or NFPA 325 *Properties of Flammable Liquids, Gases, and Solids.*

Explosions

An explosion is distinguished from a fire by its higher rate of energy release. Essentially, an explosion is the chemical reaction of a fire occurring within microseconds. An explosion typically involves:

1) a chemical reaction which produces heat (i.e., the chemistry of fire)
2) a rapid increase in pressure
3) a sudden release

Explosion damage is caused by the pressure or shock wave formed when a container holding material under increasing pressure is ruptured and there is a sudden release. For example, the Environmental Protection Agency (EPA) risk management program regulation defines the overpressure endpoint (that location where a vapor cloud explosion will do significant harm) as 1 psi.[6] This overpressure level is sufficient to partially demolish most houses.[7]

There are several special types of explosions. For example, there are boiling-liquid, expanding-vapor explosions (BLEVE) and vapor cloud explosions. ASSE's dictionary provides an example of a BLEVE as "an explosion caused when a fire impinges on the shell of a bulk liquid petroleum gas storage tank above the liquid level, causing loss of strength of the metal and consequently an explosive rupture of the tank from internal pressure."[8] Think of a BLEVE as the result of shaking and opening a warm bottle of champagne. The bubbles of carbon dioxide rising through the champagne simulate the vapor rising through the boiling liquid. The opening in the bottle is too small to fully release the expanding vapor. If the glass were to fail like the metal tank shell, a small BLEVE could occur.

Despite many theoretical and empirical studies of explosions, many of their characteristics are not well understood. When designing for safety, engineers and designers need to use reference sources carefully (for example NFPA 68, *Venting of Deflagrations* and NFPA 69, *Explosion Prevention Systems*) and apply appropriate safety margins (see Chapter 3). In addition to engineering controls for fire prevention like closed material handling systems and local exhaust ventilation, apply explosion mitigation measures such as blowout panels and other applicable over-pressure protection measures.

Flammable and Combustible Liquids

Flammable and combustible materials are the single most common fuel source involved in fires and explosions. Twelve of the 18 OSHA part 1910 hazardous material regulations are concerned with storing or handling flammable/combustible gases and liquids. Organizations frequently draft their own standards for flammable combustible materials used in their facilities that meet or exceed regulatory requirements.

When vapors from flammable liquids mix with air in sufficient concentrations to reach a critical range, or when dangerous mixtures of vapor and powder form, these mixtures can release tremendous energy. Depending on material characteristics, quantities and concentrations, confinement of the mixtures, and other factors, an ignition source can trigger an energy release ranging from a small flash fire to a huge explosion.

Fuels and oxidizers may be gases, liquids, or solids. Example fuels are:

- Gases—acetylene, butane, carbon monoxide, propane, and hydrogen
- Organic liquids—acetone, ether, gasoline, alcohols, and pentane
- Solids—paper, cardboard, plastics, flour, and magnesium particles

Oxidizer examples include:

- Gases—oxygen, fluorine, and chlorine
- Liquids—nitric acid and hydrogen peroxide
- Solids—ammonium nitrate and metal peroxides[9]

When the fuel is solid or liquid, remember that the vapors above the fuel are much easier to ignite than the fuel itself.

Flash Point

The flash point is the lowest temperature of a liquid at which it gives off sufficient vapor to form an ignitable mixture with the air near the surface of the liquid or within the vessel used. The flash point can be determined by using the closed cup method (commonly used to determine the classification of liquids that flash in the ordinary temperature range), or the open cup method (which usually gives a somewhat higher flash point).

The flash point is one of the most important physical properties that is used to determine the relative fire and explosion hazards of liquids. Comparing the flash points of gasoline (-45 degrees F) with heptane (24.8 degrees F) explains why the latter is used when training first responders how to use fire extinguishers.

Classification of Flammable Liquids[10]

The NFPA classification system is important because storage and use requirements are based on the fire characteristics of flammable and combustible materials. Flammable or class I liquids include any liquid "having a closed cup flash point below 100 degree F (37.8 degrees C), and having a vapor pressure not exceeding 40 psia (2068 mm Hg) at 100 deg. F (37.8 deg. C). Class I liquids shall be subdivided as follows:

Class IA shall include those having flash points below 73 degrees F (22.8 degrees C) and having a boiling point below 100 degrees F (37.8 deg. C)." Examples include acetaldehyde, ethylamine, ethyl ether and pentane. The purpose of the IA classification is to include all liquids whose flash points are exceeded at room temperature and that are likely to boil at ambient temperature.

"Class IB shall include those having flash points below 73 degrees F (22.8 deg. C) and having a boiling point at or above 100 degrees F (37.8 deg. C)." Examples include common solvents like acetone, cyclohexane, ethyl acetate, ethyl alcohol, and methyl ethyl ketone. This class is to distinguish materials from IA liquids that will not boil at room temperature.

"Class IC shall include those having flash points at or above 73 degrees F (22.8 deg. C) and below 100 degrees F (37.8 deg. C)." This covers flammables like amyl acetate, dibutyl ether and butyl alcohol that reach their flash point just above normal room temperatures.

Classification of Combustible Liquids[11]

Combustible liquids are those "having a closed cup flash point at or above 100 degrees F. (37.8 degrees C)." They are further subdivided into Class II and Class III categories as follows:

• "Class II liquids shall include those having flash points at or above 100 degrees F (37.8 deg. C) and below 140 degrees F (60 degrees C)." The purpose of this classification is to call out liquids that require moderate heating to reach their flash point.

• Class III liquids have flash points at or above 140 degrees and are further divided into two subclasses:

   ❑ "Class IIIA liquids shall include those having flash points at or above 140 degrees F (60 deg. C) and below 200 degrees F (93 degrees C)." Here the purpose for this temperature range is to include liquids that require significant heating (such as an external fire) to reach the flash point.

   ❑ "Class IIIB liquids shall include those having flash points at or above 200 degrees F (93 deg. C)." This classification is meant to identify liquids that may contribute to an external fire; for example what likely would happen if class IIIB materials were stored within a burning building.

Note that when a combustible liquid is heated for use to within 30 degrees F (16.7 degrees C) of its flash point, it shall be handled in accordance with the requirements for the next lower class of liquids.

Flammability Limits

The lower flammability limit (LFL) is the minimum concentration of vapor (or gas) in air (or oxygen) at which explosion or propagation of flame occurs in the presence of a source of ignition. The upper flammability limit (UFL) is the maximum concentration of vapor (or gas) in air (or oxygen) at which explosion or propagation of flame occurs in the presence of a source of ignition. In other words, vapors explode only when their concentration is between the minimum and maximum flammability limits. If the concentration of vapor is too low or too high, propagation of flame does not occur. The explosive or flammable limits are usually expressed in terms of percentage by volume of vapor or gas in air.

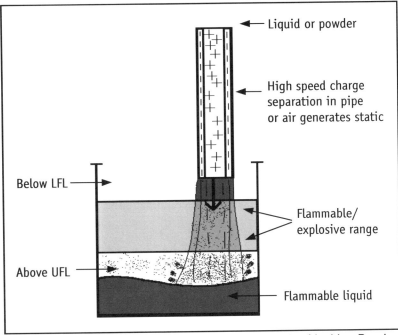

Figure 2-1 Splash Filling a Tank with a Static-Generating Liquid or Powder

This drawing illustrates a static-producing situation that could lead to a flash fire or explosion. The mixture will only ignite and burn within the flammable range. It may be useful to recall the behavior of an automobile carburetor. If the gasoline-air mixture is too lean (below the lower flammable limit (LFL)), the mixture will not ignite. On the other hand, if the mixture is too rich (above the upper flammable limit (UFL)) it will also not ignite.

In Figure 2-1, the concentration of flammable vapor outside or high in the vessel is below the LFL (too lean). Close to the liquid surface, the liquid or powder-vapor mixture has displaced enough air so that it is above the UFL (too rich). This produces a dangerous situation because it means that the flammable/explosive range is somewhere above the liquid surface. If charge separation builds up to the critical point, electrostatic discharge (a spark) may trigger a small fire or a catastrophic event.

Ignition Temperature

The Ignition Temperature is the minimum temperature to which a substance in air must be heated in order to initiate or cause self-sustaining combustion independent of the heating source. This is sometimes called the auto ignition temperature (AIT).

Ignition temperature is important since it represents the fixed temperature above which a flammable or combustible liquid will re-ignite without a spark or other ignition source. To return to the example of fire extinguisher training, assume the fuel is floating on water in a low, open top metal tank. Once the flames are extinguished, the first trainee does not have to worry about re-ignition. As the metal heats up, however, the last person in line may have put the fire out several times before the metal cools below the fuel ignition temperature. This re-ignition phenomenon is extremely hazardous to fire fighters working around large quantities of flammable liquids that have been heated by an external fire.

Vapor Density

Vapor density relative to air is important because it predicts whether the vapor will rise or fall, once released. All of the class IA vapors listed are heavier than air.

Water may be ineffective for all of the class IA liquids listed. This is particularly true for ethyl ether and pentane. Both will float on top of the water, and continued burning is likely. In most cases, alcohol-based foam is recommended since it will trap the burning vapor and cut off the air supply, thus extinguishing the flames.

NFPA Classification and Labeling of Flammable Liquids

Fire hazard properties of important organic liquids are provided in Table 2-2. The characteristics include lower and upper flammable limits, threshold limit value, IDLH concentration, and density relative to air.

The right hand column of Table 2-2 provides the NFPA 704 health, flammability, and reactivity classifications. The color and number-coded NFPA labeling system is simple and particularly useful for emergency responders that may not be familiar with chemicals handled within the plant. The colors and their meanings are:

Table 2-2 Table of Fire Hazard Properties of Flammable Liquids[12]

Flammable Liquid	Flash Point Deg. F	Ignition Temp. Deg. F	Flam. Limits % by vol. LFL	Flam. Limits % by vol. UFL	Spec. Grav. water=1	Vapor Dens. air=1	Boil. Point Deg. F	Water Soluble	Extinguishing Method	HAZARD IDENTIFICATION Health	HAZARD IDENTIFICATION Flammability	HAZARD IDENTIFICATION Reactivity
Class IA												
Acetaldehyde[1] CH_3CHO Acetic aldehyde Ethanal	-36	347	4.0	60	0.8	1.5	70	Yes	Water may be ineffective. Alcohol foam.	2	4	2
Ethylamine[2] $C_2H_5NH_2$ (Amino ethane)	<0	725	3.5	14.0	0.8	1.6	62	Yes	Water may be ineffective. Alcohol foam.	3	4	0
Ethyl Ether[3] $C_2H_5OC_2H_5$ (Diethyl ether) (Diethyl oxide) (Ethyl oxide)	-49	320	1.9	36	0.7	2.6	95	Slight	Water may be ineffective. Alcohol foam.	2	4	0
Pentane[4] $CH_3(CH2)_3CH_3$	<-40	500	1.5	7.8	0.6	2.5	97	No	Water may be ineffective.	1	4	0
Class IB												
Acetone[5] CH_3COCH_3 (Dimethyl ketone) (2-Propanone)	0	869	2.6 / 2.4 @212 F	12.8	0.8	2.0	134	Yes	Water may be ineffective. Alcohol foam.	1	3	0

Table 2-2 Table of Fire Hazard Properties of Flammable Liquids[12] (continued)

Flammable Liquid	Flash Point Deg. F	Ignition Temp. Deg. F	Flam. Limits % by vol. LFL	Flam. Limits % by vol. UFL	Spec. Grav. water=1	Vapor Dens. air=1	Boil. Point Deg. F	Water Soluble	Extinguishing Method	HAZARD IDENTIFICATION Health	HAZARD IDENTIFICATION Flamma- bility	HAZARD IDENTIFICATION Reactivity
Carbon Disulfide[6] CS_2 (Carbon bisulfide)	-22	194	1.3	50	1.3	2.6	115	No	Water may be ineffective except as a blanket	2	3	0
Cyclohexane[7] C_6H_{12} (Hexahydrobenzene) (Hexamethylene)	-4	473	1.3	8	0.8	2.9	179	No	Water may be ineffective.	1	3	0
Ethyl Acetate[8] $CH_3COOC_2H_5$ (Acetic ester) (Acetic ether) (Ethyl ethanoate)	24	800	2.2	110	0.9	3.0	171	Slight	Water may be ineffective. Alcohol foam.	1	3	0
Ethyl Alcohol[9] C_2H_5OH (Grain alcohol) (Cologne spirits) (Ethanol)	55	689	3.3	19	0.8	1.6	173	Yes	Water may be ineffective. Alcohol foam.	0	3	0
Hexane[10] $CH_3(CH_2)_4CH_3$ (Hexyl hydride)	-7	437	1.1	7.5	0.7	3.0	156	No	Water may be ineffective.	1	3	0

Table 2-2 Table of Fire Hazard Properties of Flammable Liquids[12] (continued)

Flammable Liquid	Flash Point	Ignition Temp.	Flam. Limits % by vol. LFL	Flam. Limits % by vol. UFL	Spec. Grav. water=1	Vapor Dens. air=1	Boil. Point Deg. F	Water Soluble	Extinguishing Method	Health	Flamma-bility	Reactivity
Isopropyl Alcohol[11] $(CH_3)_2CHOH$ (Isopropanol) (Dimethyl carbinol)	53	750	2.0	12	0.8	2.1	181	Yes	Water may be ineffective. Alcohol foam.	1	3	0
Methyl Alcohol[12] CH_3OH (Methanol) (Wood alcohol) (Columbian spirits)	52	725	6.7	36	0.8	1.1	147	Yes	Water may be ineffective. Alcohol foam.	1	3	0
Methyl Ethyl Ketone[13] $C_2H_5COCH_3$ (2-Butanone) (Ethyl methyl ketone)	21	960	1.8	10	0.8	2.5	176	Yes	Water may be ineffective. Alcohol foam.	1	3	0
Class IC												
Amyl Acetate[14] $CH_3COOC_5H_{11}$ (1-Pentanol Acetate)	77	680	1.1	7.5	0.9	4.5	300	Slight	Water may be ineffective. Alcohol foam.	1	3	0
Butyl Alcohol[15] $CH_3(CH_2)_2CH_2OH$ (Butanol-1) (Propylcarbinol) (Propyl methanol)	84	689	1.4	11.2	0.8	2.6	243	Yes	Water may be ineffective. Alcohol foam.	1	3	0

HAZARD IDENTIFICATION

Table 2-2 Table of Fire Hazard Properties of Flammable Liquids[12] (continued)

Flammable Liquid	Flash Point Deg. F	Ignition Temp. Deg. F	Flam. Limits % by vol.		Spec. Grav. water=1	Vapor Dens. air=1	Boil. Point Deg. F	Water Soluble	Extinguishing Method	HAZARD IDENTIFICATION		
			LFL	UFL						Health	Flamma-ability	Reactivity
Dibutyl Ether[16] (C$_4$H$_9$)$_2$O (1-Butoxybutane) (Butyl ether)	77	382	1.5	7.6	0.8	4.5	286	No	Water may be ineffective. Alcohol foam.	2	3	0
Nitroethane[17] C$_2$H$_5$NO$_2$	82	778	3.4		1.1	2.6	237	Slight	Water may be ineffective except as a blanket. Alcohol foam. Explodes on heating.	1	3	3
Turpentine[18]	95	488	0.8		<1		300	No	Water may be ineffective.	1	3	0
Xylene-o[19] C$_6$H$_4$(CH$_3$)$_2$	90	869	1.0	6.0	0.9	3.7	292	No	Water may be ineffective.	2	3	0

Note: Flash Point, Ignition Temp., and Boiling Point are given in deg. F.
Sp. Gr. of water = 1.0;
Vapor Density of air = 1.0.

(1) Polmerizes. See hazardous chemicals data.
(2) 70% aqueous solution
(3) Underwriter Lab Class 100. See hazardous chemicals data.
(4) CH3(CH2)3CH3
(5) Underwriter Lab Class 90. See hazardous chemicals data.
(6) Underwriter Lab Class 110+. See hazardous chemicals data.
(7) Underwriter Lab Class 90-95. See hazardous chemicals data.
(8) Underwriter Lab Class 85-90.
(9) Underwriter Lab Class 70.
(10) Underwriter Lab Class 90-95.

(11) Underwriter Lab Class 70. See hazardous chemicals data.
(12) Underwriter Lab Class 90. See hazardous chemicals data.
(13) Underwriter Lab Class 85-90.
(14) Underwriter Lab Class 55-60. See hazardous chemicals data.
(15) Underwriter Lab Class 40. See hazardous chemicals data.
(16) Dibutyl ether
(17) See hazardous chemicals data.
(18) Underwriter Lab Class 40-50.
(19) Underwriter Lab Class 40-45. See hazardous chemicals data.

- Blue—for health hazard information such as toxicity or corrosiveness
- Red—indicates the material's relative flammability
- Yellow—relates to the material's instability or reactivity
- White—for special hazards

The numbers run from zero to four and indicate the degree of hazard based on the color category.

- Zero—the lowest hazard and safest. For example, zero in the red area means the material is non-flammable. In the blue area, zero means the material is non-toxic or non-corrosive to the skin and eyes.
- Four—Highest hazard and most dangerous. For example, four in the red area (acetaldehyde, ethylamine, ethyl ether and pentane) means the material is highly flammable. Note that class IB and IC flammable liquids have a NFPA flammability rating of 3.

The NFPA rating system should be used for chemicals throughout the plant. Give all employees the opportunity to identify the meanings of the colors and numbers during hazard communications training.

Every facility storing or handling chemicals that will burn needs a copy of NFPA 30—Flammable and Combustible Liquids Code. NFPA 704, "Identification of Materials," is also a useful resource, as is NFPA 325, "Fire Hazard Properties of Flammable Liquids, Gases and Volatile Solids," and NFPA 49, "Hazardous Chemicals Data." NFPA information is also reported on many material safety data sheets meeting OSHA's Hazard Communication standard.

Comparing Liquid Class and Fire Hazard Ratings

Attempts to achieve a consistent classification and labeling system for flammable liquids has proven to be an elusive project for many years. For example, experts have attempted to develop some kind of consistent criteria across the following four hazardous material situations for years:[13]

1. Transportation and labeling (Department of Transportation definitions, UN Model Regulations, and others)
2. Waste disposal (EPA "Characteristics of Ignitability")
3. Storage and handling (OSHA, NFPA, API, and others)
4. Emergency response (NFPA 704)

Table 2-3 NFPA 30 Classification System for Liquids with Corresponding NFPA 704 Fire Hazard Ratings[14]

Liquid Class	Hazard Rating	Criteria (deg. C)	Criteria (deg. F)
IA	4*	BP < 37.8 and FP < 22.8	BP < 100 and FP < 73
IB	3	BP ≥ 37.8 and FP < 22.8	BP ≥ 100 and FP < 73
IC	3	22.8 ≤ FP < 37.8	73 ≤ FP <100
II	2	37.8 ≤ FP < 60	100 ≤ FP < 140
IIIA	2	60 ≤ FP < 93.4	140 ≤ FP < 200
IIIB	1**	FP ≥ 93.4	FP ≥ 200
0	0	5 minute IT > 815.5	5 minute IT > 1500

BP Boiling point. For single component liquids the boiling point is defined as the temperature at which the vapor pressure is equal to 760 mm Hg. Fore mixtures that do not have a constant boiling point, the 20% evaporation point of a distillation performed in accordance with ASTM 86 *Standard Method of Test for Distillation of Petroleum Products* is considered to be the boiling point.

FP Closed-cup flash point. NFPA 30 Chapter 1-7.4 defines the method to be used according to liquid viscosity with some exceptions being permitted.

IT Ignition Temperature. A material that will not burn (fire hazard rating = 0) is defined in NFPA 704 as not burning in air after 5 minutes exposure to a temperature of 2500 deg. F (815.5 deg. C). The test method for IT is not given in JFPA 704. From the discussion in the 1969 edition of NFPA 325M, IT is based on some type of auto ignition temperature test. ASTM's current auto ignition test is ASTM E 659 *Standard Method for Auto ignition Temperature of Liquid Chemicals.* However, an auto ignition test may identify liquids as being unable to burn if they contain volatile, non-combustible components; this is because the vapor-air mixture in the test flask may be rendered non-combustible as these components evaporate. DOT has adopted an ad hoc procedure for removing volatile, less flammable materials prior to conducting a flash point test. Such a procedure could be used to avoid the preceding problem in the IT test.

* Fire hazard rating of 4 is also given to flammable gases, flammable cryogenic materials, CS_2, and materials that ignite spontaneously when exposed to air, regardless of material state.

** Per NFPA 704 (1996), this fire hazard rating is also given to "Liquids with a flash point greater than 95 deg. F (35 deg. C) that do not sustain combustion when tested using the *Method of Testing for Sustained Combustibility,* per 49 CFR Part 173 Appendix H, or the UN *Recommendations on the Transport of Dangerous Good,* 8th Revised Edition." Also given to "Liquids with a flash point greater than 95 deg. F (35 deg. C) in a water-miscible solution or dispersion with a water non-combustible liquid/solid content of more than 85% by weight." Also given to "Liquids that have no fire point when tested by ASTM D 92, *Standard Test Method for Flash point and Fire Point by Cleveland Open Cup,* up to the boiling point of the liquid or up to a temperature at which the sample being tested shows an obvious physical change."

There are significant differences between DOT and NFPA systems for transportation and storage/handling/labeling.

Table 2-3 provides the liquid class, hazard rating, and liquid classification criteria for NFPA 30 and NFPA 704.

Table 2-4 DOT Criteria for Class 3 Flammable Liquids[15]

Packing Group	Criteria (deg. C)	Criteria (deg. F)
I.	IBP ≤ 35 deg. C	IBP ≤ 95 deg. F
II.	IBP > 35 and FP < 23 deg. C	IBP > 95 and FP < 73 deg. F
III.	FP ≥ 23 deg. C	FP ≥ 73 deg. F
	IBP > 35 and FP ≤ 60.5 deg. C	IBP > 95 and FP ≤ 141 deg. F

IBP: initial boiling point

Exceptions:

1. Liquids meeting criteria for Class 2 Flammable gas

2. Liquid mixtures containing at least 99 volume % of components having flash points ≥ 60.5 deg. C (141 deg. F) provided the mixtures are not offered for transportation or transported above their flash points

3. Liquids with FP > 35 deg. C (95 deg F) that fail the sustained combustion test given in 49 CFR Part 173 Appendix H

4. Liquids with FP > 35 deg. C (95 deg. F) and fire point > 100 deg. C (212 deg. F) according to ISO 2592

5. Any liquid with FP > 35 deg. C (95 deg F) which is in a water miscible solution with a water content > 90 weight percent

In the U.S., NFPA 30 classifies flammable and combustible liquids based on flash points and boiling points. The NFPA system is referenced in several NFPA Fire Codes and is particularly applicable to emergency response situations. Liquids with a flash point below 100 degrees F are classified as flammable.

Transportation situations involve risks not normally present in stationary storage and handling facilities. For example, increased possibility of mechanical damage, variable environmental impacts along the route, lack of continuous fire protection, and longer gaps in emergency response time all represent greater risks in transporting flammable liquids along rail and roadways. Further, in rail cars or trucks, liquids can easily attain temperatures approaching 140 degrees F without deliberate heating. Thus DOT classifies flammable liquids as having flash points less than or equal to 141 degrees F.

A third labeling system (ANSI Z129.1) from the American National Standards Institute differs again from NFPA and DOT systems.

The point is of these comparisons is to make you aware of the regulatory complexity for shipping and handling chemicals; this will help enable you to implement procedures that meet the requirements.

Flammable Liquid Storage and Handling Requirements

NFPA 30, *The Flammable and Combustible Liquids Code,* is the recognized authority on storing, loading, and unloading flammable liquids. Origin of the standard goes back to 1913 when it was written in the form of a municipal ordinance. Development of NFPA 30 has continued since those early days and now provides detailed information as follows:[16]

Chapter
1. General Provisions
2. Tank Storage
3. Piping Systems
4. Container and Portable Tank Storage
5. Operations
6. Referenced Publications

Appendix
A. Explanatory Material
B. Emergency Relief Venting for Fire Exposure for Aboveground Tanks
C. Abandonment or Removal of Underground Tanks
D. Suggested Fire Protection for Containers of Flammable and Combustible Liquids
E. Fire Test Data
F. Fugitive Emissions Calculations
G. Referenced Publications

A copy of NFPA 30 is a must for those designing systems and defining policy and procedures involving flammable and combustible liquids. Table 2-5 is taken from Section 2-5 of the code to help users locate storage tank buildings based on tank capacity, stable vs. unstable liquids, and level of emergency relief provided.

Table 2-5 Location of Storage Tank Buildings with Respect to Property Lines, Public Ways, and the Nearest Important Building on the Same Property[17]

Largest tank- operating liquid capacity (gal)	Minimum distance from property line that is or can be built upon, including opposite side of public way (ft)				Minimum distance from nearest side of any public way or from nearest important building on same property (ft)			
	Stable liquid		Unstable liquid		Stable liquid		Unstable liquid	
	Emergency Relief		Emergency Relief		Emergency Relief		Emergency Relief	
	Not over 2.5 psig	Over 2.5 psig	Not over 2.5 psig	Over 2.5 psig	Not over 2.5 psig	Over 2.5 psig	Not over 2.5 psig	Over 2.5 psig
Up to 12,000	15	25	40	60	5	10	15	20
12,001 to 30,000	20	30	50	80	5	10	15	20
30,001 to 50,000	30	45	75	120	10	15	25	40
50,001 to 100,000	50	75	125	200	15	25	40	60

SI Units: 1 gal = 3.8 L; 1 ft = 0.3 m; 1 psig = 6.9 kPa.

Double all distances shown if protection for exposures is not provided. Distances need not exceed 300 ft.
Capacity of any individual tank shall not exceed 100,000 gal without the approval of the authority having jurisdiction.

Electrostatic Discharge

Ignition of flammable vapors or combustible dusts in a confined space like a vessel or building is often the source for devastating fires and explosions. Electrostatic discharge (ESD) is frequently the ignition source. Flammable limit information, such as that provided in Table 2-2, is critical to our understanding of both hazards and mitigation measures involving flammable vapors. Lower and upper flammable limits must be considered when dealing with vapor-air mixtures resulting from plant operations like splash filling (see Figure 2-1).

As we have seen from the example of splash filling above, the hazards of ESD involve generation, accumulation, and discharge in a flammable atmosphere. A critical factor here is that people having no experience with ESD do not anticipate the nature of this unseen hazard.

Examples of ESD Hazards

Many solids in particle form are severe ESD hazards because they generate and accumulate static during transport, handling and processing. Physical properties of the powder are a factor in static generation; however, the processing method has a much greater effect. Dangerous operations for solids include:

- Sliding on a chute or scoop shovel
- Impact against walls during pneumatic conveying
- Grinding operations

Similar hazardous processing for non-conductive liquids involve:

- Spraying
- Long free-fall distances and splashing
- High-shear blending and emulsifying
- Fine filtering operations

Eliminating and Neutralizing Electrostatic Discharge

The usual methods for ESD control include analysis of solids/powders and liquids for electrostatic discharge properties, analysis of the process and equipment, bonding and grounding, humidification (not effective for high speed processes or closed systems), as well as process design and installation. Safe process design considerations should include:

- Removing the flammable atmosphere
- Substituting or adding high conductivity liquids
- Anti-static additives
- Reducing liquid flow velocity
- Dip pipe filling
- Choice of conductive contact materials for construction
- Provide bonding, grounding, and verification
- Use of static relaxation chambers for receiving vessels
- Special precautions involving agitation and atomization of flammable liquids
- Minimizing fine filtration
- Methods for reducing electrostatic charge generation and buildup
- Special precautions such as inerting vessels for transferring solids and powders

EPA Risk Management Program (RMP)

OSHA's Process Safety Management regulation and EPA's "Chemical Accident Prevention Provisions" (RMP regulation) include coverage for flammable substances and the prevention/mitigation of and response to fires and explosions. See Appendix A and B for an overview of the EPA RMP Guidance Document and the use of computerized release and dispersion models. This information is used to meet EPA's requirement to predict offsite consequences of potential fires and explosions.

Fire and Explosion Safety

The National Fire Protection Association (NFPA) and Chapter 6, "Fire Protection," of the National Safety Council's *Accident Prevention Manual, Engineering and Technology*[19] are the two comprehensive resources for fire and explosion safety that go well beyond the scope of this chapter. Purchase appropriate NFPA standards and the set of three APM reference books for your facility library.

Good judgment must be used in establishing an appropriate level of fire protection. Start with a broad scope, considering the following factors and resources:

- Prior incident history
- Process hazards and process hazards analysis (PHA) studies
- Government regulations

Table 2-6 Hazardous Location Classifications[22]

Class I Highly flammable gases or vapors		Class II Combustible dusts		Class III Combustible fibers or flyings	
Division 1	Division 2	Division 1	Division 2	Division 1	Division 2
Locations where hazardous concentrations are probable or where accidental occurrence should be simultaneous with failure of electrical equipment	Locations where flammable concentrations are possible, but only in the event of process closures, rupture, ventilation failure, etc	Locations where hazardous concentrations are probable, where their existence would be simultaneous with electrical equipment failure, or where electrically conducting dusts are involved	Locations where hazardous concentrations are not likely, but where deposits of the dust might interfere with heat dissipation from electrical equipment, or be ignited by electrical equipment	Locations in which easily ignitable fibers or materials producing combustible flyings are handled, manufactured, or used	Locations in which such fibers of flyings are stored or handled except in the process of manufacture

Group Classifications:

Class I

A – Atmospheres containing acetylene

B – Atmospheres containing hydrogen or gases or vapors of equivalent hazard

C – Atmospheres containing ethyl-ether vapors, ethylene, or cyclopropane

D – Atmospheres containing gasoline, hexane, naphtha, benzene, butane, propane, alcohol, acetone, benzol, or natural gas

Class II

E – Atmospheres containing metal dust, including aluminum, magnesium, and other metals of equally hazardous characteristics

F – Atmospheres containing carbon black, coke, or coal dust

G – Atmospheres containing combustible dusts not included in Group E or F, including flour, grain, wood, plastic, and chemicals

- Corporate and facility guidelines and policies
- Public resources (including local fire department and organizations like the NFPA and American Petroleum Institute)
- Plant emergency response capability
- Outside consultant recommendations[20]

OSHA regulations related to fire safety are found in 29 CFR 1910 Subparts H (Hazardous Materials) and L (Fire Protection). Note that OSHA incorporates NFPA standards by reference, so look to those standards as guidelines to compliance.

Facility safety coordinators may contact the local fire department, city/state office of public safety, or city/state fire marshal for an annual inspection.

This type of inspection by an outside expert fits into the scope of inspections discussed in Chapter 5. Ask inspection teams to look for ignition sources and other fire hazards during plant-wide and department self-inspections.

As discussed above, flammable gases or vapors, combustible dusts, and combustible fibers are a dangerous source of fuel in most facilities. Be sure to self-inspect or call in an expert to assure proper hazardous location classifications (Table 2-6).

Emergency Planning and Response

Facility safety and health professionals need to act as internal consultants to help their management through the process of identifying potential emergencies and developing primary and contingency response plans.

Prevention of personal injury and loss of life is the prime objective of emergency planning. Step one is to define and document emergency situations into some kind of so-called "Red Book" or Emergency Manual. Then recruit volunteers for an emergency squad, assign responsibilities, outline procedures, and establish a set of drills appropriate to potential scenarios.

Types of Emergencies

Identifying the types of emergencies that can affect the facility will facilitate the planning, organizing, and implementing of effective response plans. These include:

- Fire and explosions
- Medical emergencies (serious injuries including fatalities)
- Chemical overexposures
- Hazardous material releases and spills
- Utility interruptions

- Floods
- Hurricanes and tornadoes
- Earthquakes
- Other natural disasters
- Civil strife, sabotage, and terrorist threats
- Site and community security
- Workplace violence
- Work accidents and rumors
- Emergency shutdowns
- Wartime emergency management
- Other emergencies or security issues specific to the location

The hazards from these emergencies will vary somewhat depending on the facility location, production and engineering processes, and work practices.

Emergency Squads, Training and Drills

Emergency planning and response is certainly not an exact science. However, facilities need to do the planning at a level that will quickly allow a practical response once the emergency has quickly been assessed. Because response time is at a premium, facility management should rely on facility first responders rather than off-site resources. OSHA calls out the requirements for a fire brigade in 29 CFR 1910.156. Because these requirements are quite stringent, a facility may be better off to establish an *Emergency Squad* (short hand is E Squad or EQ) and systematically train these people to handle a variety of emergencies. As the learning and equipment improve, the organization may wish to make the EQ an official fire brigade. Training for fire brigades requires:

- Quarterly sessions for those responding to interior structural fires
- A higher level of training for employees of the oil refining industry
- Information about potential exposures, such as storage and use of flammable and toxic materials
- Written procedures provided by the employer

Fire drills should be carefully planned and practiced at least annually. Instruction sheets that include evacuation routes need to be posted and kept up to date. Note that employees need to evacuate immediately (even in the event of a very small fire) unless they have been trained in the use of fire extinguishers. Fire extinguisher training is designed to teach employees how to stop fires from spreading out of control. Provide the right extinguishers (type A, B, or all-purpose) so that they are immediately reachable and may be promptly used.

Disaster Planning and Management

Once management has done some advance planning and has evaluated the types of emergencies that could affect the facility, it is time to begin an action plan. In most cases, this will require coordination with local emergency planning agencies in order to protect facility people, business operation, and the community. The following program considerations should be undertaken while preparing an emergency manual:[23]

1. Company policy, purposes, authority, principal control measures, and emergency organization chart showing positions and functions
2. Some description of the expected disasters with a risk statement
3. Map of the plant, office, or store showing equipment, medical and first aid, fire control apparatus, shelters, command center, evacuation routes, and assembly areas
4. List and post cooperating agencies along with contact information
5. Plant warning systems
6. Central communications center, including home contact information for employees
7. Shutdown procedure, including security measures
8. How to handle visitors and customers on-site
9. How to handle other locally related and necessary matters
10. A list of equipment and resources that are available and how to initiate action

Other plan of action considerations include: chain of command, training, hazardous waste and spills (HAZMAT), command headquarters, emergency equipment, alarm systems, facility protection and security, emergency medical services, and transportation.

Outside help topics related to emergency planning and response are: mutual aid plans, contracting for disaster services, municipal fire and police departments, industry and medical agencies, as well as other governmental and community agency assistance.

Appoint and train several teams of people to assist with potential disasters. Give consideration to: a fire brigade, emergency response team, spill control team, hazardous materials incident team, emergency medical re-

Figure 2-2 Typical Duties of an Emergency Response Team[24]

- Select and use various types of fire extinguishers effectively
- Provide first aid and CPR
- Protect themselves and others from potential consequences of blood borne pathogens
- Conduct shutdown procedures
- Control and mitigate chemical spills
- Select and use respirators and self-contained breathing apparatus (SCBA) as needed
- Conduct search procedures for potential victims
- Conduct emergency rescue operations (including confined space rescue)
- Respond to hazardous material emergencies in accordance with 29 CFR 1910.120.

sponse team, evacuation wardens, rescue team, search team, and salvage team. Figure 2-2 lists duties of a typical emergency response team.

Take advantage of chemical site security work that has been accelerated since September 11, 2001. The American Chemistry Council, Synthetic Organic Chemical Manufacturers Association, and the Chlorine Institute have cooperated to prepare the "Guide to Site Security in the US Chemical Industry."[26]

Government Hazardous Material Response Teams

Be sure to check out what services are offered by federal, state and local response teams. Minnesota has one of the best-equipped state hazardous materials organizations in the country. Their network includes:[25]

- An emergency response mitigation team in the capital, St. Paul;
- Ten chemical assessment teams in South Anoka County, Duluth, Hopkins, northern Minnesota Iron Range, Mankato, Moorhead, Rochester, St. Cloud, St. Paul, and Willmar;
- Four police bomb squads under state contract;
- Hazardous materials technicians and specialists at fire departments in Bemidji, Bloomington, Carver County, Cottage Grove, International Falls, Marshall/Luverne, Minneapolis, Plymouth/Maple Grove, Winona and elsewhere;
- Most fire departments have trained first responders who can secure a scene and rescue victims;
- State agencies having hazardous material response capabilities include: Agriculture, Bureau of Criminal Apprehension, Emergency Man-

agement, Fire Marshall, Health, Natural Resources, Pipeline Safety, Pollution Control, Transportation;

- Federal agencies: Bureau of Alcohol, Tobacco, and Firearms; Drug Enforcement Administration; Federal Bureau of Investigation; and
- National Guard: 55th Civil Support Team in the Twin Cities plus a Duluth unit.

Check to see what public emergency planning and response services your location may provide. The most critical aspect of EP&R is to coordinate facility and public resources.

Conclusion

In order to help the reader integrate the information in this chapter into a chemical safety management program, we provide a few concluding points.

Start by asking the facility safety committee to address that most useful question "What and where are the hazards?" You want to identify process sections and other areas of the facility having fuels and potential ignition sources that could lead to a fire or explosion. Ask, "Do we have any of the following situations:

- Open containers or locations for flammable vapors?
- Nearby ignition sources?
- Locations or procedures where electrostatic discharge could cause a fire or explosion?
- Storage of wood, cardboard, paper and other combustible materials near flammable/combustible liquid storage?
- Uncontrolled cutting, welding or other hot work activities by employees and contractors?
- Potential sources for internal/external release of flammable/combustible chemicals?

Next, apply the fire and explosion safety measures described in this chapter. Review the provisions of NFPA 30, paying particular attention to applicable storage and piping systems and operations within the facility. Study the site plan for location of flammable/combustible storage tanks and buildings with respect to property lines, public ways and nearby buildings. Review chapter 6 on fire protection in NSC's Accident Prevention Manual, Engineering and Technology.

Establish fire protection policies and procedures that meet or exceed regulatory requirements (Chapter 10). Set up a 3-ring binder of EHS policies and procedures that includes fire safety. Many facilities separate fire related items covering policies like electrical classification, hot work, portable fire extinguishers, fixed fire protection systems, fire and emergency exits and shelters, etc. into a separate binder often called the "Red Book." Put in place a family of self-inspections and inspections/audits by the local fire marshal and other experts (Chapter 5).

Finally, provide fire and emergency response training and exercises for the facility emergency squad. Involve the local fire department to promote their knowledge of the facility and the potential for mutual aid.

References

1. Reprinted with permission from "Recognizing the causes of fire," © 2000 Factory Mutual Insurance Company, P.O. Box 7500, Johnston, RI 02919.

2. *Accident Prevention Manual for Business & Industry, Engineering and Technology, 12th Edition*, p.141, National Safety Council, 1998.

3. *Accident Prevention Manual for Business & Industry, Engineering and Technology, 12th Edition*, p.142, National Safety Council, 1998.

4. 3M trademark, 3M, St. Paul, Minnesota.

5. *Accident Prevention Manual for Business & Industry, Engineering and Technology, 12th Edition*, p.142, National Safety Council, 1998.

6. Risk Management Program Guidance For Offsite Consequence Analysis, p. 1-8, Chemical Emergency Preparedness and Prevention Office, U.S. Environmental Protection Agency, www.epa.gov/ceppo, April 1999.

7. V.J. Clancey, "Diagnostic Features of Explosion Damage." Sixth International Meeting of Forensic Sciences, Edinburgh (1972).

8. *Dictionary of Terms Used in the Safety Profession, Third Edition*, American Society of Safety Engineers, 1988.

9. List adapted from *National Electrical Code 1993*, NFPA 70, p. 466-467.

10. Definitions from NFPA 30 Flammable and Combustible Liquids Code, 1993 Edition, ANSI/NFPA 30, National Fire Protection Association, p. 30-8.

11. Definitions taken from Flammable and Combustible Liquids Code, 1993 Edition, ANSI/NFPA 30, National Fire Protection Association, p. 30-8.

12. Selections reprinted with permission from NFPA 325M *Flammable Liquid, Gases, Volatile Solids* Copyright ©1969, National Fire Protection Association, Quincy, MA 02269. This reprinted material is not the complete and official position of the National Fire Protection Association, on the referenced subject which is represented only by the standard in its entirety.

13. Laurence G. Britton, Survey of Fire Hazard Classification Systems for Liquids, *Process Safety Progress,* Vol.18, No.4, winter 1999. Reproduced with permission of the American Institute of Chemical Engineers. Copyright ©1999 AIChE. All rights reserved.

14. Laurence G. Britton, Survey of Fire Hazard Classification Systems for Liquids, *Process Safety Progress,* Vol.18, No.4, winter 1999. Reproduced with permission of the American Institute of Chemical Engineers. Copyright ©1999 AIChE. All rights reserved.

15. Laurence G. Britton, Survey of Fire Hazard Classification Systems for Liquids, *Process Safety Progress,* Vol.18, No.4, winter 1999. Reproduced with permission of the American Institute of Chemical Engineers. Copyright ©1999 AIChE. All rights reserved.

16. NFPA 30 *Fire Hazard Properties of Flammable Liquids, Gases, Volatile Solids*, National Fire Protection Association, Quincy, MA, 1969.

17. Reprinted with permission from NFPA 30 *Flammable and Combustible Liquids Code* Copyright ©1993, National Fire Protection Association, Quincy, MA 02269. This reprinted material is not the complete and official position of the National Fire Protection Association, on the referenced subject which is represented only by the standard in its entirety.

19. *Accident Prevention Manual for Business & Industry, Engineering and Technology, 12th Edition*, p.133, National Safety Council, 1998. .

20. Donald L. Roe, "Fire Protection Risk Assessment: A Proposed Methodology for Oil and Chemical Facilities," *Process Safety Progress,* (Vol. 19, No. 2), Summer 2000.

21. Reprinted with permission, Accident Prevention Manual for Business & Industry, Engineering & Technology, p. 135, 12th edition, National *Safety Council,1998.*

22. Adapted from Article 500, NFPA 70 National Electrical Code® Copyright © 1992, National Fire Protection Association, Quincy, MA 02269. This material is not the complete and official position of the National Fire Protection Association, on the referenced subject which is represented only by the standard in its entirety.

23. *Accident Prevention Manual for Business & Industry, Administration and Programs, 12th Edition*, p.156, National Safety Council, 1998.,

24. Reprinted with permission, Jack E. Daugherty, Industrial Safety Management, A Practical Approach, Government Institutes, ABS Group, Rockville, Maryland, p. 641, 1999.

25. Minnesota Department of Public Safety, 2001.

26. http://www.socma.org or http://www.americanchemistry.com

Chemical Process Design Considerations

Introduction

Good engineering project management requires assurance that hazards and risks have been identified early in the design stage and that adequate capital and other resources are available to address these potential problems. There are several objectives associated with the goal of integrating safety and design:

- Minimize exposure to employees and to the public
- Protect facility assets and continuity of operation
- Comply with federal and local regulations
- Comply with company and facility standards and guidelines

Identifying chemical hazards and associated risks requires process knowledge and documentation. Process, project, and design engineers need to work together to build facilities and equipment.

Many chemical process design considerations have an enormous "upfront" effect on safety and health. That is, design decisions made at early stages of engineering projects have a great impact on personnel and process safety in the plant.

Introduction to Safety in Design

Before designers can plan to make things work right, they must think about what can go wrong. Consider the hazard analysis component of EPA's Risk Management Planning regulation in which the facility is required to describe and prevent potential worst-case and alternative scenarios that involve covered chemical releases.

For many facilities, thousands of release scenarios are possible. Thus, it is not practical to detail every scenario, but we can use EPA guidelines to develop example scenarios that fit our facility. Include failures such as pipe ruptures, holes in storage tanks and process vessels, ground spills, etc., that meet the EPA worst-case and alternative criteria. Figure 3-1 provides examples of the largest predictable (alternative scenario) and largest potential (worst-case scenario) releases.

Figure 3-1 Examples of Worst-Case and Alternate Chemical Releases

Worst-Case Release: one that results in greatest distance to toxic endpoint, or a vapor cloud explosion overpressure of one pound per square inch.

- Rupture of one or more storage tanks containing toxic or flammable substances
- Overturned rail car or truck trailer containing toxic or flammable materials
- Pipe rupture or large flange leak from a major process line (> 4" diameter)
- Transfer hose release due to splits or sudden hose uncoupling near ignition source and other vessels/tanks containing toxic and/or flammable materials

Alternate Release: one that is more likely to occur than the worst case release and that will reach an endpoint offsite, unless no such scenario exists.

- Failure of process piping at flanges, joints, welds, valves, etc.
- Manual overfilling of storage tanks or situations with automatic shutoff
- Overpressure and venting to atmosphere from relief valves and rupture disks
- Breakage or puncture of shipping containers

Function of Codes and Standards

Codes and standards are extremely important in developing engineering specifications and making design decisions. Hazards addressed and precautions taken as a result of a regulatory or company policy provide structure and boundaries for the specifications and the design.

OSHA's comprehensive Process Safety Management standard, 29 CFR 1910.119, has fourteen elements in a format that literally defines a plant's compliance program. Two of the elements apply directly to the design function:

1. Process Safety Information (PSI)
2. Mechanical Integrity (MI)

PSI and MI are critical inputs to project engineers and designers because this information concerns process and equipment knowledge.

Process Safety Information

Those that drafted the OSHA PSM standard recognized the importance of documenting strong chemical, process, and equipment understanding. Consider the list of requirements for process safety information codified under 29 CFR 1910.119(d).

Chemical Hazard Information

- Toxicity information for input chemicals, intermediates, and products
- Permissible exposure limits for regulated chemicals
- Physical, reactivity, and corrosive data for all chemicals
- Thermal and chemical stability data for all chemicals
- Hazardous effects due to inadvertent mixing of different materials

Technology of the Process

- Block-flow diagram or simplified process-flow diagram
- Process chemistry and material balance(s)
- Maximum intended inventory
- Safe upper and lower limits for temperatures, pressures, flows, and/or compositions
- Evaluation of the consequences of deviations, including those affecting safety and health

Equipment in the process

- Materials of construction
- Piping and instrument diagrams (P&IDs)
- Electrical classification
- Relief system design and design basis
- Ventilation system design
- Design codes and standards employed
- Material and energy balances
- Safety systems like interlocks, detection systems, and suppression systems

Project and design engineers need to use this required process safety information in such a way that it will function as an equipment specification.

Chemical Hazard Information

Some of the required chemical safety information comes from the material safety data sheets (MSDS) required by OSHA's Hazard Communication Standard (29 CFR 1910.1200(g)) and supplied by the material manufacturer or distributor. MSDS information is the very least of the data needed by designers. Note that the quality of MSDS information for the same chemical is highly variable, depending on the source.

Further, most MSDSs cover pure components and do not speak to reactivity and hazard information on mixtures. Process and equipment designers need more than this generic chemical safety information. In many cases, they also need actual laboratory scale up data for the system.

The Data Package

The following is a good example of the kind of information found in a complete chemical data package:[1]

- Chemical identification data, such as chemical formula and synonyms
- Physical properties data, such as molecular weight, density, boiling point, freezing point, vapor pressure, viscosity, and solubility
- Thermodynamic data, such as reaction heat, latent heat, heat capacity, and thermal conductivity
- Reactivity/stability data, such as flashpoint, flammability limits (or minimum explosive concentration and ignition energy for dusts), decomposition or polymerization (uninhibited monomers), shock sensitivity, pyrophoric nature, reactivity with common materials of construction, and trace quantities of other chemicals
- Exposure guidelines such as threshold limit value, immediate dangerous to life and health value, and emergency response planning guideline levels
- Regulatory status data, such as OSHA-permissible exposure limits, EPA SARA reporting quantities, DOT classification and EPA RCRA classification
- Toxicity data such as LC (50), LD (50), carcinogenicity, mutagenicity, teratogenicity, and neurotoxicity data
- Special requirements for raw materials, including H_2O or nitrogen, for upset conditions
- Time-temperature sensitivity of mixtures and in-process streams.

MSDS information is often the best source of data. However, research and development personnel, suppliers, plant safety coordinators, project engineers, and designers often need to work together to generate relevant information that may not be provided by the MSDS. Documentation is critical. The laboratory, engineering, and manufacturing organizations need a design information system that is current and readily accessible to authorized personnel.

As the data package is assembled, designers need to ask these and other guiding questions:

- Do we have material and energy balances for all reactions?
- Are there side reactions that can generate toxic or explosive conditions, clog critical equipment, or cause other dangerous conditions?
- Have proper materials of construction been specified?
- Have adequately-sized pressure relief devices been provided for normal and emergency situations?
- Does the storage arrangement address material incompatibilities?
- Does the design account for the possibility of contaminants to upset the process? Are there possibilities of cross contamination from one process to another?
- Are containers sized to minimize personnel exposure?
- Has spill control been provided for loading and unloading areas?
- Can large quantities of hazardous materials be moved easily in the event of an emergency?

Other technical sources for important safety information on materials include:

- *Dangerous Properties of Industrial Materials* (Sax), Van Nostrand Reinhold and other toxicology references provided in Chapter 1.
- *Perry's Chemical Engineer's Handbook*, McGraw-Hill
- *The Merck Index*, Merck and Co.
- *Handbook of Chemistry and Physics*, CRC Press
- *Handbook of Industrial Hazard Assessment Techniques*, The World Bank
- *Fire Protection Handbook*, National Fire Protection Association

Note that some physical, reactivity, corrosivity, thermal, and chemical stability data is not available in handbooks. For a price, you can outsource specific tests on your materials or do the work in-house. Calorimetry is a good source for generating thermodynamic and kinetic models and data where highly exothermic and/or gas-producing reactions are concerned.[2]

Process Technology Information

From the list of OSHA process technology requirements above, one sees that this information is process-specific and not always readily available during the early design stage. For example, you should determine the maximum intended inventory of a hazardous chemical, including quantities that could be stored in multiple use containers like tote tanks. Thus the design function can provide working capacities for storage tanks, reactors, drums, and other vessels having either fixed or variable levels. To assure clarity, inventories should be expressed in terms of a measured parameter such as the output of a level detector. In this manner, plant personnel can easily compare actual and maximum intended inventories with no conversion and little or no interpretation.

Operating and Safety Process Limits

Equipment and control system designers need information relating to operational and safety limits for all process components. Much of this information will come from the initial process hazards analysis. For new equipment, use the hazard and operability (HAZOP) method, since the deviations of high pressure/temperature/flow rate/vessel liquid levels, etc. are ideal for developing the process limits required by OSHA's process safety management standard.

Note that organizations generally have normal process operating targets and product quality limits from the scale-up data package. Quality limits are tighter than operational or safety limits.

Establish four parameter limits for each process component (Figure 3-2). Provide CRT indication and protection to counteract deviation for key parameters like overpressure as follows:[3]

1. Pressure indicator-controller (PIC) at 100 psi (normal control point)
2. Sound high pressure alarm at 150 psi (quality limit)
3. Powerful high-high pressure alarm at 175 psi (operation limit)
4. Emergency shutdown system activates at 185 psi (safe limit)

Process Equipment

Material of construction information should show which materials are used and document why they are chosen. Reference a particular code or standard that specifies material versus type of service wherever possible.

Note that electrical area classification comes from NFPA 70 or the National Electrical Code NEC 50. This set of guidelines will help designers classify each area of the facility to minimize ignition sources (see Table 2-6). Most piping and instrumentation diagrams (P&IDs) should indicate electrical classification.

Much has been written on relief system design and design basis. This information provides the rationale for location and sizing of safety relief

Limits of Operation	Unit _____	Sheet ____ of ____															
Line Number	Name	Temperature, F				Pressure, PSIG				Flow Rate, gal. or lbs./min				Consequences of Deviation
		Normal range	Quality limit	Operation limit	Safe limit	Normal range	Quality limit	Operation limit	Safe limit	Normal range	Quality limit	Operation limit	Safe limit	

Figure 3-2 Operating and Safety Process Limits

valves (SRVs), pressure relief valves (PRVs), and rupture disks (RDs). You will want to establish a worksheet or report that summarizes information and calculations to specify PRV locations, inlet and discharge piping, and PRV set point.

Ventilation systems design is also part of the mechanical system. Documentation is often found on the process flow diagram (PFD) or P&ID. Again, design data such as airflow and equipment sizing calculations are normally found in the design report or other backup documents.

Safety Systems

OSHA requires facilities to list safety systems as part of the process safety-related equipment described above. This documentation should include:

- Control interlocks that automatically slow or stop operation of critical equipment until certain process parameters are within acceptable ranges. The interlocks may stop equipment that is running or prevent standby equipment from starting.
- Systems designed to de-pressure the process completely or partially.
- Disposal and containment systems for excess quantities of hazardous material. Examples are flares, scrubbers, and holding tanks.
- Suppression systems for toxic or flammable materials as they are released. Water deluge and spray systems are good examples.
- Detection systems for toxic or flammable materials. These instruments are available commercially for a number of common HAZMAT chemicals. Before choosing a device, remember that reliability may be an issue, given rugged outdoor and process environments.

Piping and Instrumentation Diagrams (P&IDs)

Along with MSDSs, P&IDs are one of the most important documents in the process safety information package. Not only is a P&ID often the most descriptive document a facility has for the process, it is often the primary document used during the process hazard analysis. P&ID documentation is the primary output of the design department. Often the safety portion of this package does not receive adequate attention.

By definition, a P&ID is a diagram representing the as-built/as-modified status of the process system design. It must be kept up to date for the life of the process (red pencil lines and hand written notes are acceptable until changes are brought up to date). The American National Standards Institute (ANSI) provides guidelines on format and symbols to be used.

However, most companies and facilities use a system that fits their individual needs. The following information is typically found on a P&ID drawing package:[4]

- All process components, including spares with an identification or tag number
- Size and specification for all piping, including instrument tubing, and the location and size of pipe reducers and expanders
- Flow directions
- Symbols along with identification of instruments, alarms, interlocks, and trip points
- All valves and fail-safe positions
- Identification of insulation and heating/cooling systems for all piping, pumps, and vessels
- Steam traps, filters, strainers, and other in-line equipment
- Elevation measurements for all equipment and important nozzle connections
- Connections for venting, draining, sampling, and flushing; size of the connection and destination (examples are flares, sewers, drain funnels, etc.)
- Mechanical utility and waste treatment connections including those to other process systems
- Identification of all electro-mechanical safety equipment and their set points
- Piping slope, critical clearances and other important notes related to safety and operability

Mechanical Integrity

In addition to maintaining accurate process safety information, the OSHA PSM standard requires employers to ensure the initial and ongoing integrity of process equipment. Mechanical integrity has a high-design component. That is, without a complete documented rational for the design, the manufacturing plant has no basis for maintaining the equipment according to the PSM code.

Early in the equipment design process, engineering designers are looking for codes—guidelines and standards that will help them accurately specify equipment. Further, equipment is often designed for multiple applications. Reactors, tanks, and other vessels are often designed for specific use and later applied to production of other chemical products and intermediates. This kind of multiple-use application makes design documentation even more important than it might seem at first glance.

Minimum requirements for mechanical integrity are codified under 29 CFR 1910.119(j). Facilities are in full compliance with the regulation if the following requirements are in place:

1. *Written procedures exist for maintaining ongoing integrity of process equipment.*

   This requirement is for pressure vessels, storage tanks, piping systems, relief and vent systems, emergency shutdown systems, controls and monitoring devices, pumps, compressors, filters, and fired heaters.

2. *The facility has established a training program for maintenance personnel.*

   Training must include an overview of the process and its hazards, as well as instruction in applicable safety procedures.

3. *Equipment inspections and testing are done for all covered processes.*

   This must be done according to recognized and generally accepted engineering practices.

   The frequency of inspections and testing is determined by applicable codes or standards (or more frequently based on prior operating experience). When test results fall outside acceptable limits, equipment deficiencies must be corrected prior to use or in a safe and timely manner to ensure safe operation.

4. *Documentation of inspection and testing of process equipment is in place.*

   Documentation includes the date of inspection, name of the inspector, serial number or other identifier of the equipment, description of the inspection or test performed, and results of the inspection.

5. *A quality assurance system for the construction of new plants and equipment is in place.*

   Equipment as fabricated must be suitable for the intended process application. The equipment needs to be installed properly and be consistent with design specifications and the manufacturer's instruction. Maintenance materials, spare parts, and equipment shall be suitable for the intended process application.

   Be aware that pressure vessels and other chemical process and storage equipment are designed, built, and maintained according to an appropriate set of rules, such as those established by the American Society of Mechanical Engineers (ASME) and by organizations like the American Petroleum Institute (API).

Other Useful Standards

Additional sources with design codes and standards relevant to process safety information and mechanical integrity for the chemical processing industry are:

- American Society of Civil Engineers
- American Institute of Chemical Engineers (AIChE)
- American National Standards Institute (ANSI)
- American Society for Testing and Materials (ASTM)
- Building Officials and Code Administrators International, Inc.
- National Fire Protection Association (NFPA)
- Hartford Steam Boiler Inspection and Insurance Company
- Institute of Electronics and Electrical Engineers (IEEE)
- Instrument Society of America
- National Association of Corrosion Engineers
- American Gas Association (AGA)
- Chlorine Institute
- Compressed Gas Association (CGA)
- Chemical Manufacturers Association (CMA)
- Tubular Exchangers Manufacturers Association (TEMA)
- American Insurance Services Group (AIS)
- Bureau of Mines
- U.S. Coast Guard/Department of Transportation (DoT)
- U.S. Environmental Protection Agency (EPA)
- U.S. Occupational Safety and Health Administration (OSHA)
- National Institute for Occupational Safety and Health (NIOSH)/ Department of Health and Human Services
- Underwriters Laboratories, Inc. (UL)
- National Electrical Code, National Building Code, and other national, state, or local codes

See Figure 8-7 for a listing of consensus organizations and applicable standards for design considerations.

Electronic Document Management

Design departments may want to purchase an electronic document management (EDM) system. Regulations like OSHA's Process Safety Management (PSM) standard are only one of the factors driving increased process complexity and time-consuming recordkeeping demands on facility. An EDM system adds management control to the document stream for important elements of PSM like process safety information and mechanical integrity.[5] EDM systems like Database Application's *CADEXnet* (which uses "object-oriented" versus older "relational" technology) link each document in the system to a database record. Using the example of design calculations for sizing a pressure relief valve, this means one could electronically retrieve the basis for sizing the valve by clicking on the corresponding symbol within the P&ID.

Safety Procedures for Capital Projects

This section covers safe design and other procedures that fit into what is commonly called the *Engineering Project Process*. The project process describes those steps starting with simply a new or expansion concept and ending in a functioning facility complete with operational process equipment. Organizations generally have their own terms for the stages of a project process, and the stages themselves may vary from company to company. Facilities need an understanding of how safety fits into each step of the project process.

One way of looking at the project process is in terms of the following five steps:

1. Conceptual stage
2. Basic engineering
3. Detailed engineering
4. Construction
5. Commissioning

It is more useful, however, to integrate safety and health into the project by breaking the project down into six stages:

1. Project scope and staffing procedures
2. Authorization for Expenditure (AFE) or request for funds procedures
3. Hazard review procedures
4. Facility and equipment siting procedures

5. Process and design review/documentation procedures

6. Completed project

Project Scope and Staffing

Correctly defining the project scope will determine the content and detail of the entire effort. Applying the appropriate level of technology, getting the right people involved, and drafting a clear proposal are the essential elements of this stage.

In this day and age of intensive competition for limited resources, the business will often benefit if the engineering leaders propose a broader scope than can be implemented in a single project. For example, engineering, marketing/product planning may meet to request approval for a 5–10 year capital facilities/equipment improvement program integrated with the long-range business plan. Once the overall plan is modified and approved, the engineering team can confidently proceed to provide the detailed scope of a smaller project, knowing that incremental capital will be released by top management as projects consistent with the business plan are approved. Putting a long range proposal together can be a tricky and frustrating process, but the rewards are great.

Protection of employees, the community and the environment needs to be thoroughly considered at the front end of a project. Here we are concerned with potential fires, explosions, toxic releases, and employee exposures to other hazards.

The American Institute of Chemical Engineer's Center for Chemical Process Safety (CCPS) recommends a screening mechanism to assist in determining the degree of risk and the level of funding necessary to address the risks involved.[6]

- *"High safety risk*—a project incorporating technology or processes that are either new or unproven to the company or that involve hazardous materials, toxics, flammables, etc., of sufficient quantity and potential energy to endanger employees, customers, or third parties.
- *Medium safety risk*—a project incorporating proven/established company technology that has not received a quantitative risk and consequence analysis but is judged to have a low probability of endangering employees, customers, or third parties from hazardous events.
- *Low safety risk*—a project involving the reapplication of technology already familiar to the company for which risks have previously been

evaluated. The location or operation does not significantly affect customers or third parties."

Project managers may not be able to categorize potential projects in the medium to low risk levels unless the organization has the safety infrastructure in place to have previous evaluations completed.

Staff the project team according to the project scope. Choose a project engineer/manager and include staff and plant process/control engineering, plant engineering/maintenance functions; plus staff and plant safety/health/environmental coordinators. A common mistake in staffing is not to involve process instrumentation and control specialists early in the project such that information and control system strategies are not appropriate to the level of hazard. Reserve monies for specialists in chemistry, toxicology, industrial hygiene, environmental engineering, and others that might be needed as the project proceeds, but take advantage of the expertise you need when developing project scope. Consult early in the project scope stage with a corporate risk evaluation expert or seek outside help.

There are some special considerations unique to chemical and related industries at this stage. With increased public and governmental concerns about chemical hazards, organizations will do well to carefully monitor on a long-range basis, the safety, health, transportation and environmental regulatory climate before making large commitments. For example, a common mistake here is to design and build large scale or multiple plants without adequate allowance for increasing restrictions on air emissions and more demanding resources to prevent all chemical releases.

Authorization for Expenditure (AFE)

AFE is a term commonly used to describe the procedure for requesting approval to spend capital funds. The important thing about this part of the project process is to thoroughly define the project scope before writing the final proposal. Be sure to categorize and answer questions about projected safety/health and environmental expenditures for such things as process instrumentation and control, area electrical classification, ventilation and blow down discharge systems, material transfer and storage systems, waste collection and disposal systems, special maintenance systems and equipment, etc. in order to simplify later writing and tracking of job and shop work orders. Build safety and project review milestones into the project timeline like design reviews and process hazard analysis.

Study and follow the organization's management system or protocol for approval and release of capital funds. Remember that senior management approving project requests usually prefer an opportunity to speak with someone in person before approving the AFE document. Be prepared to walk the paperwork through the approval cycle rather than relying on the mail system. Track progress of the approval document and remain on-call, so that you are immediately accessible by phone or in person when the approving executive wants to talk about the project.

Hazard Reviews

The goal here is to minimize risk in a cost-effective manner. Thus the management system and procedures for hazard reviews must assure that required process safety information is summarized in the AFE document. This means that the chemical hazard, technology of the process and process equipment information described in the first part of this chapter, must be condensed into a paragraph or two. As stated above, identify the hazard reviews in the project time table. Process and safety review procedures must be integrated into the entire project system, from start to finish. The process hazards analysis (PHA) is the first step in hazard review sequence.

Facility and Equipment Siting

Generating this information will start with the initial process hazards analysis (PHA). OSHA requires the PHA study team to consider facility/equipment siting and make a qualitative evaluation of possible safety and health effects of failure of controls on employees. The EPA goes further to define parameters required for modeling the release and dispersion of toxic and flammable materials that could have an offsite impact (Appendix A). Hazard assessment software is used to quantify the offsite impact of a release (Appendix B).

The well-known story of the Bhopal, India accident (Chapter 4) is a useful lesson learned relative to plant siting, although location was not an issue when first built.

Superimpose a release and dispersion model on the plot plan, summarize the information in the project proposal, and make it a permanent part of the project file. While making the offsite impact assessment for hazardous materials, also consider potential exposures and hazards to and from adjacent plants facilities. Also factor-in the potential impact of natural disasters like earthquakes, floods, tornados, etc.

In addition to process and storage locations, the detailed siting analysis needs to include loading/unloading areas and utility locations. Note that a feature of some PHA software is an impact analysis of the loss of plant utilities and services.[7]

Process and Design Review/Documentation

The purpose of this set of procedures is to facilitate process understanding and to assure that a quality design is implemented correctly. A strong, "robust" process depends on thorough process understanding on which to base the design.

The project team needs to document the design intent for each process section along with detailing independent/dependent variable relationships and associated process safety information.

Hold design reviews that include process safety items before and after the design package is sent out for bids and construction contracts awarded. Here, the focus should be on verifying that corrective action has been taken to address action items generated during the PHA.

Quality assurance procedures are directed to using the right materials of construction, proper fabrication and inspection methods, and that equipment installation addresses concerns identified at the site. For example, procedures for installation of pressure relief devices, including use of appropriate gasket material needs to be verified. It is a good idea to conduct an initial walkthrough and periodic supplier audits to ensure equipment purchases are suitable for the intended application. Management of change procedures are used where needed to authorize facility/equipment changes from the original design.

Documentation records for the reviews should leave an understandable "paper trail" that verify the design basis has been faithfully carried out. Establish a simple system for addressing process and design team recommendations:

1. Go: implement the recommendation as documented.
2. No-go: project management believes the recommendation is not justified. In this case, the item should be reviewed separately with EHS hazard specialists and documentation included that supports the final decision.
3. Alternative: There is agreement with the intent of the recommendation but project management has an alternative. Alternatives must

also be reviewed by EHS functions and supporting documentation included.

Project management's final report completes the project process as well and declares that, following a final inspection, the equipment is ready for commissioning to the facility. The report needs to list each review team recommendation and how it was handled. Include references to all supporting documentation.

Completed Project

A project is complete when the process equipment is fully operational and checked out. This is the commissioning stage and means *the wall* in the facility is taken down and the new system is formally turned over to production. This transition is virtually seamless, if all the project procedures above have been followed.

Safe Design and Engineering Culture

In order to consistently integrate safety fully into a design function, a change in engineering culture is often needed. Engineers must realize that the total cost of a project includes those costs associated with preventable injuries incurred, not just costs of preventive measures. To understand this point, consider two basic principles for integrating safety into design:[8]

1. Safety should be continuously considered through all phases of design.
2. Process discipline and structure are needed to help the designer focus on safety because many design professionals have limited training in safety engineering and hazard avoidance.

To counteract the thinking that safety can be addressed "afterwards," the safety and design team must exercise project process discipline and direction as discussed above.

Designers are encouraged to draft safety fitness-for-use (FFUs) criteria[8] that address hazards such as:

- Removing or controlling high levels or unusual forms of energy
- Complying with regulatory requirements based on obsolete technology
- Providing user-friendly ergonomics for procedures to replace manual material handling such as transferring solids to a vessel

- What percentage of design engineers view safety as a post-design consideration?

- Is instilling safety in a design viewed as an added cost or simply as a characteristic of good design?

- On how many capital projects are safety FFUs provided? Design reviews conducted? Safety walk-downs completed?

- What project cost overruns were attributed to correcting safety hazards? How many dollars per year?

- What percentage of capital projects had plant involvement in design safety reviews?

- How many engineering leaders have included design safety goals in the performance objectives for their engineers?

- How many engineers have had performance evaluations positively or negatively affected due to their safety performance on designs?

- What rewards exist for designers who consistently produce safe designs?

- What processes are in place for providing feedback to design engineers on field modifications to their designs for safety reasons?

- How many hours of training do engineers receive annually on safety-related aspects of design?

- How have the average number of safety punch items changed over time?

- What percentage of projects were delayed or missed the scheduled start-up date due to work required to correct safety hazards?

- How many safety specialists maintain schedules of capital projects for their facilities and routinely track progress?

- What processes have been established to audit design safety review and walk-down quality?

- How many facilities operate in countries for which the engineering organization does not have current safety standards?

- When designing systems for foreign operations, what standards and guidelines are followed if local standards are weak or non-existent?

- How much weight is safety given in due diligence studies for potential acquisitions?

- What safety requirements are included in contracts for engineering services? Generic safety specifications? Training and proficiency requirements?

- What processes exist for auditing safety performance of outside engineering service providers?

- How are external providers held accountable for safety design deficiencies—who absorbs the costs?

Figure 3-3 Benchmarking the Safe Design Process[9]

- Controlling noise such as that coming from a high-speed mixer
- Making effective changes on nearby processes outside the project scope

It is useful to benchmark or "compare notes" with design cultures within or outside the organization that are further along in making this continuous improvement change.

Figure 3-3 provides a series of questions to help determine how well design safety is currently integrated into an organizational culture. The questions can be used to compare current and historical performance, across various internal business units, and involving outside organizations.

Design Strategies and Examples

The internal/external project and design engineering groups need to fully develop process and instrumentation drawings and specifications so that the facility can maintain the original design intent for the equipment in each process section.

The regulatory basis for chemical safety design is OSHA's mechanical integrity requirement for process safety management (1910.119(j)). This section of the code calls out specific requirements for:

1. Mechanical integrity of pressure vessels and storage tanks, piping systems including valves, relief and vent systems/devices, emergency shutdown systems, process controls including monitoring devices, sensors, alarms and interlocks; and pumps.
2. Written procedures for maintaining ongoing mechanical integrity
3. A facility training program for maintenance people
4. Inspection and testing of process equipment
5. A system for addressing equipment deficiencies
6. A quality assurance system for construction of new plants and equipment.

The core requirement here involves number 4, equipment inspections and tests that satisfy the American Society of Mechanical Engineers (ASME) consensus standard for boilers and pressure vessels.[10] Facilities and equipment engineering organizations need to keep originals or copies of documents that include each manufacturer's data report, as-built drawings, code calculations, nameplate rubbings or photos, certified material test reports, hydrostatic test reports of time versus pressure and other documents required by the ASME code.

The mechanical integrity PSM code has become a driving force for change in chemical and related process industries. For the years 1992–1994, mechanical integrity citations are ranked number two among PSM violations (Chapter 8). Many engineering groups are going back to gather the needed documentation described above for covered equipment and control systems. Facilities are increasingly focused on inspections/tests and appropriate frequencies.

OSHA requires that inspections/tests be consistent with applicable manufacturers' recommendations and good engineering practices, and more frequently based on prior operating experience. Manufacturers are often not told enough about the equipment application to recommend an inspection/test frequency. Facilities are prone to replace or clean safety devices like pressure relief valves only when a malfunction occurs. For example, a plugged relief valve may be discovered during each annual turnaround inspection for several years. Devices like this should be removed and tested more frequently until operations and maintenance discover how often the plugging occurs.

Intrinsically Safe Designs

AIChE summarizes four basic strategies for design and operation of inherently safe processes developed by safety expert Trevor Kletz and accepted in the engineering community.[11]

1. **Intensification/Minimization**—minimize the quantity of hazardous substances. This is generally accomplished by reducing inventory or using, for example, smaller and/or continuous reactors to contain hazardous chemicals.

   Early in his career, the author worked to safely scale-up a process to make and purify a solid high energy oxidizer used in the formulation of solid rocket propellants. The tiny metal reactors contained approximately one quart of the reactants. Purification of the oxidizer was done in glassware in order to prevent a "missile" from damaging other equipment in the event of an explosion. Both reactor and purification processes were contained in separate rooms, having one-foot thick poured, steel reinforced concrete walls and remotely controlled from an adjacent "control room." A thick Plexiglas™ window allowed the operators to see the process. Manually operated valve stems were extended through the wall as were cables to operate other electro-mechanical controls. A flimsy "blowout door" was provided to safety relieve a small explosion out of the operating bay into a large fenced part of the arsenal grounds. The limit for direct handling of oxidizer samples was 0.25 grams of solid material. Personal protective gear consisted of a thick leather jacket (worn backwards and tied in back) and gloves, a full- face mask over safety glasses and special hearing protection.

Minimizing quantities processed and handled in this way also minimized the effects of occasional explosions of the highly sensitive material. Time lost was generally the only consequence and this too, was minimized by having substitute equipment on wheels and ready to move into the operating bay once the mess had been cleaned up.

2. An example of **substitution** is replacing a material with a less hazardous one.

3. **Moderation** is a third way to make a process inherently safer. Use less hazardous conditions, materials, or facilities that minimize the impact of a release of hazardous material or energy. Here an example is operating a distillation column filled with flammable liquid and vapor under vacuum in order to moderate processing pressures and temperatures.

4. **Simplification**—design to eliminate unnecessary complexity and make operating errors less likely. Continuous chemical processing, with simple in-line mixers is an example of eliminating un-needed steps often involved in batch operations.

Equipment Overpressure Protection

There were 758 mechanical integrity violations found by OSHA during the period 1992-1997. Many of these had to do with improperly designed or installed overpressure protection. A study of pressure relief systems involving over 250 units, indicates that "nearly half of the equipment in the oil, gas, and chemical industries lack adequate overpressure protection as defined by recognized and generally accepted good engineering practice."[12] The study presents data collected from a large number of pressure relief system design audits conducted by an independent contractor. The audits were mandated by OSHA under the Process Safety Management (PSM) regulation to determine how effective process hazards analysis (PHA) studies were in evaluating pressure relief systems.

The study shows that 59% of the overpressure systems surveyed met the PSM standards. This means 41% did not meet the standard. The breakdown of the 41% is as follows:

No relief device—15%

Undersized device—7%

Improper installation—17%

Undersized and improperly installed device—2%

Unfortunately, nearly all of the systems were "designed by (and installed under the supervision of) reputable design firms." The authors of this paper, presented at the Loss Prevention Symposium held during the AIChE Spring National Meeting in Atlanta, Georgia on March 7, 2000, "conclude that, as a practical matter, conventional PHA methods are ineffective tools for evaluation pressure relief systems." In this author's opinion, it is not so much that the PHA tool is ineffective as that too often the study teams lack the expertise and do not take the time necessary to study adverse consequences of overpressure deviations.

PHA teams need to be multi-disciplined, involving engineering designers and manufacturing people with experience in the process under analysis. Engineering files need to include code calculations, checked by the equipment manufacturer and approved by safety and the project manager. Design teams need references that provide safe criteria for overprotection. These include API guidelines[13] and section VIII of the ASME Boiler and Pressure Vessel Code.[14]

In addition, scenarios should be studied that involve potential causes for high pressure. Include high temperature, high levels downstream, blocked lines and vents, control valve failure, heat exchanger failure, thermal expansion of cold liquids, buildup of non-condensible vapors and external fire. Finally consider "penthouse" type enclosures to contain extremely toxic materials that may be released to the atmosphere through rupture disks, pressure relief valves, and "blowdown" tanks.

Design of Process Control Systems

It is easy to overlook the OSHA mechanical integrity requirement covering process controls. Facilities are required to maintain the original design intent for monitoring devices, sensors, alarms and interlocks. Years ago, the standard for safety-related process instrumentation and control systems required that they be "hard-wired"— not part of process control software.

Safety instrumented systems (SIS) are modern replacements for hard-wiring safety controls independent of the distributed process controls.

In a broad-reaching and practical article that summarizes "lessons learned" from safety instrumented systems design, Paul Gruhn presents the findings of a study in the U.K. that reviewed 34 accidents and found the following causes of control and safety system failure:[15]

1. Specifications—44%
2. Changes after commissioning—20%
3. Operations and maintenance—15%
4. Design and implementation—15%
5. Installation and commissioning—6%

Well over half of the accidents were caused "where technical decisions were involved, such as deciding what and how well the system should perform." Because others have found similar problems, there has been an increasing emphasis on industry standards for process control. The documents referenced come from Europe and key U.S. consensus standards:

- *Guidelines for Safe Automation of Chemical Processes*[16]
- *Application of Safety Instrumented Systems for the Process Industries*[17]

Safety instrumented systems (SIS) is only one of many layers of protection surrounding a facility (see Figure 3-4) and separating the process from people and the environment.

The basic process control system should always be the primary safeguard. Management depends on critical alarms and operator intervention (as a secondary safeguard) to prevent incidents where for some reason, the control system does not maintain the process within safe limits. Safety instrumented systems (SISs) is but the third level of protection from the

---

- Community Emergency Response
- Plant Emergency Response
- Physical Protection (Containment Dikes)
- Physical Protection (Relief Devices)
- Safety Instrumented Systems
- Critical Alarms Operator Intervention
- Basic Process Control Systems

Figure 3-4 Multiple Independent Safety Layers[18]

process, activated when something in the process goes beyond alarm limits. Additional protective measures would need to fail in order for the impact of a chemical release, fire, or explosion to have offsite consequences. This layered approach to safety is what the nuclear industry calls "defense in depth."

Conclusion

Essential chemical design considerations are driven by the need to develop and maintain process safety information and system mechanical integrity system for the life of the process. A detailed chemical and process technology data package is critical to developing accurate facility and equipment specifications. If the data is incomplete, the information must be generated and documented prior to releasing the final design package for bids.

Operating and safe upper and lower limits must be established for important process variables. Descriptions of facilities and equipment are needed, including materials of construction, piping and instrumentation diagrams, electrical classifications, material and energy balances, and all safety systems and devices used.

Finally, there are six key points for the successful design of chemical process systems:

1. Establish a rigorous engineering project process and set of procedures.
2. Employ appropriate design codes and standards to generate facility and equipment specifications. Draft internal standards that meet or exceed regulatory requirements.
3. Pay special attention to overpressure situations that can lead to loss of containment for hazardous materials.
4. Engineering and manufacturing organizations need to adopt an iterative "Safety Life Cycle" approach to the design of safety instrumented systems:[19]
   a. Analyze the hazards with the goal of inherent safety.
   b. Determine the performance required in terms of acceptable risk.
   c. Choose a technology, level of redundancy, and methods for testing the new system.
   d. Evaluate the system reliability and measure performance against the original target.

5. Involve process instrumentation and control functions early in the process. These people are the last ones out of the plant before startup. Therefore, they should be among the first to participate at the beginning of the project.

6. Train project engineers (include hands-on laboratory sessions) in the following process control fundamentals: basic elements of control strategies and systems, principles of instrumentation and feedback, and use of personal computers for process control.

References

1. *Plant Guidelines for Technical Management of Chemical Process Safety*, CCPS, p. 34.

2. Richard C. Wedlich, Chilworth Technology, Inc., "Tame Process Thermal Hazards Using Calorimetry," *Chemical Engineering Progress*, September, 1999, p. 71.

3. Adapted from Comelio De la Cruz-Guerra and M. Javier Cruz-Gomez, "Using Operating and Safety Limits to Create Safety Procedures," *Process Safety Progress* (Vol. 21, No. 2), American Institute of Chemical Engineers, June 2002, Table 1.

4. *Chemical Process Safety Report*, Thompson Publishing Group, Inc., June 1997, p. 11.

5. http://www.cadexnet.com

6. *Plant Guidelines for Technical Management of Chemical Process Safety*, Center For Chemical Process Safety of the American Institute Of Chemical Engineers, New York, New York, 1992, p. 48.

7. Hazard Review Leader™, ABS Consulting, www.jbfa.com.

8. Paul S. Adams, "Establishing a Safe Design Process," *Professional Safety*, American Society of Safety Engineers, November, 1999.

9. Paul S. Adams, "Establishing a Safe Design Process," *Professional Safety*, American Society of Safety Engineers, November 1999. Reprinted with permission.

10. ASME Boiler and Pressure Vessel Code, ASME International, www.asme.org.

11. Safety and Health News, A Supplement to Process Safety Progress, American Institute of Chemical Engineers, Spring 2000 and *Process Plants: A Handbook for Inherently Safer Design*, T.A. Kletz, Taylor & Francis, Bristol, Pa (1998).

12. Patrick C. Berwanger, Robert A. Kreder and Wai-Shan Lee, "Analysis Identifies Deficiencies in Existing Pressure Relief Systems," *Process Safety Progress,* (Vol. 19, No.3), fall, 2000.

13. "Guide for Pressure-Relieving and Depressuring Systems", API Recommended Practice, 521 Fourth Edition (March 1997), The American Petroleum Institute, www.api.org.

14. ASME Boiler and Pressure Vessel Code, ASME International, www.asme.org.

15. Paul Gruhn, "Safety Instrumented System Design: Lessons Learned," *Process Safety Progress,* Vol. 18, No. 2, fall 1999, Figure 1. Original study is "Out of Control, Why Control Systems Go Wrong and How to Prevent Failure," *UK Health & Safety Executive,* Sheffield, England (1995).

16. Publications, ISBN No. 0-8169-0554-1, www.aiche.org, American Institute of Chemical Engineers, Center for Chemical. Process Safety, New York, NY (1993).

17. ANSI/ISA S84.01, ISBN: 1556175906, The Instrumentation Systems and Automation Society, www.techstreet.com, RTP, NC (1996).

18. Adapted from Paul Gruhn, "Safety Instrumented System Design: Lessons Learned," Figure 3. Material used with permission of AIChE, Vol. 18, No. 2, Fall 1999.

19. Paul Gruhn, "Safety Instrumented System Design: Lessons Learned," *Process Safety Progress,* page 157-158, Vol. 18, No. 2, fall 1999.

Accident and Incident Investigation

Introduction

Effective accident and incident investigation is a means to prevent similar events from happening again. A good accident-incident investigation establishes all relevant facts about how the accident occurred, as well as why it occurred. Furthermore, how a facility handles information gained from unplanned incidents is an excellent measure of the facility's entire chemical safety management system.

Because *all* accidents are preventable, all accidents need to be investigated. This philosophy extends to investigation of incidents (generally regarded as accidents without injuries), near accidents (sometimes called close calls or near misses), and potential hazards (accidents waiting to happen).

We know that accidents can be very costly, in both direct and indirect terms. Direct costs (medical expenses plus hours lost times the burden rate) are insured. However, workers compensation premiums are based on a formula using a history of actual claims. Therefore, two similar plants may pay annual premiums ranging from tens of thousands of dollars to six or even seven-figure amounts, depending on their relative injury and illness experience.

Note that indirect costs (facility and equipment damage, business interruption, product scrap/rework, additional training costs, and other hard-to-measure expenses) may range from four to ten times direct costs. Thus, total costs for an injury with one day lost and minor medical expenses may jump from a few hundred dollars to several thousand. To gain management attention to the real costs of an accident, calculate the *added sales* necessary at the company's average profit level to make up for the accident. It is not unusual to find that the added sales may range from 40 to 80 times the original direct costs! This crude "cost of an accident" analysis adds a compelling motivation to accident and incident investigation.

There are many other benefits to a good accident investigation program. Accident investigation promotes continuous improvement in a facility's chemical safety and health infrastructure. Because the investigation often involves a multi-disciplinary team, employee involvement and team-building also results.

Introduction to Accident and Incident Investigation

Hazards are the immediate causes of accidents, either in the form of unsafe work conditions or unsafe work practices. Most often, a combination of unsafe conditions and practices are at the core of an accident-incident.

All accidents, incidents, and near misses, including the following, need to be investigated:

- Major incidents such as fire, explosions, runaway reactions, or releases of toxic or flammable materials
- Accidents with injuries
- Accidents without injuries (e.g. chemical releases and spills)
- Near accidents and close calls
- Potential hazards

Problems Faced by Safety Coordinators

There can be many mistakes made during an accident investigation. However, we can learn from a number of common errors and omissions that plants experience repeatedly:

- The investigation is not taken far enough to identify root causes.
- The investigation is used to assign blame.
- The cause is attributed to unsafe work conditions without consideration of procedures, work practices, or behaviors.
- An unsafe act or failure to act is assumed to be the cause of an accident without regard to underlying management system problems.
- Facts are confused with conclusions.
- Corrective actions are recommended before a full investigation is complete.
- Accident scene observations, interviews, and documentation review are incomplete, inadequate, or not done at all.
- Investigation reports are incomplete, missing facts, poorly supported, or misstated; conclusions, assumptions, and opinions are stated as

facts; and the writing style is poor (choice of words, too brief, too many words, lack of continuity, inflammatory wording, and/or editorializing).

- There are "organizational misconceptions" (concept proposed by B.A. Turner) that lead to an accident. These may be true misconceptions, misinformation, rejection of information, or problems delivering information to different levels in a company.[1] The misconceptions may be reinforced by economics, misapplication of technology, and/or inadequate assessment of risk. They may or may not be recognized in the accident investigation report.

The first two problems above—investigations that are not taken far enough or investigations that are used to assign blame—are, by far, the most common. Because the manufacturing culture is action-oriented, it is often tempting to stop the investigation once the immediate cause is determined and move on.

Often, the accident cause is blamed on an unsafe condition. Note that a phrase such as "faulty or defective equipment" in a report is a red flag to upper management and OSHA. This style of reporting can result in a reprimand crashing down on the facility manager's head. Sometimes, co-workers or the union may use "faulty or defective equipment" as an excuse to absolve the worker of any blame.

When an unsafe act or failure to act is identified, blame often clouds the investigation process. If the failure to act is assumed to be the cause of the accident, then the system problems (reasons for failure to act) may not be revealed.

A supervisor or manager may label the accident as caused by "employee error." Identify potential human error factors in the examples below.

Levels of Accident Investigation

There are at least two levels of accidents and incidents based on the scope of damage and on the degree of complexity involved. Personnel-related accidents may be of simple or high complexity.

The following story is an example of a simple personnel-type accident.

A worker is splashed in the eye while diluting sulfuric acid by pouring water into the acid. We might conclude that safety glasses, or personal protective equipment (PPE), could have prevented the eye injury; thus lack of PPE is the cause. However, PPE would not have prevented the splash. So this accident is not as simple as it at first seems. Lack of PPE is the *immediate* or a *contributing cause* to the accident. However, the basic or *root cause(s)* of the accident have to do with failure to follow a simple rule of thumb when diluting corrosive chemicals with water. The rule is to always gently pour the chemical into the water, not the other way around.

Note that there are other important cause-related questions that indicate gaps in the plant's safety management system, including "Was there a written procedure?" and "Was the person trained to pour the acid correctly?" If the answer is no, we have lack of training and/or a lack of a written procedure to add to our list of potential causes.

Simple accident investigations should always be conducted by the area supervisor or manager. He or she may call on the site safety coordinator for input.

The next example is concerned with a more complex personnel-related incident.

An employee in an area next to process equipment transfers toluene out of a 55-gallon drum into a large glass beaker. He is in a hurry to get this task completed and is not thinking of the hazards involved in handling a flammable liquid. Consider the potential for static buildup between the toluene surface in the beaker and the drum due to the flowing liquid (Figure 2-1). The farther the liquid falls, the more the static charge will build. Furthermore, the worker is not wearing PPE (gloves and safety glasses with side shields and/or an appropriate face shield). He does not use a bonding wire and metal probe between the drum and the glass beaker to equalize the electrical potential. Suddenly a static discharge occurs within the vapors above the toluene in the beaker.

Because the concentration of toluene vapors in air is above the LFL of 1.2% by volume and below the UFL of 7.1%, the mixture ignites. The worker, startled by the flames, drops the beaker. In his panic, he does not immediately let go of the safety faucet on the drum so that toluene continues to spill on the floor. The flames spread from the burning liquid on the floor to other nearby containers of flammable and combustible liquids. Because some of the containers are plastic, they quickly soften from the heat and release their contents. The worker's cotton pant legs and nylon smock catch fire and the flames accelerate as he runs from the area and past a safety shower. Before long, half of the building is in flames.

Hearing the commotion, someone pulls the fire alarm. The sprinkler system discharges over the flames, but because of the quantity of flammable and combustible organic materials, the burning continues until the fire department arrives ten minutes later.

The worker received third degree burns and 20 other employees in the building were evacuated without injury. There is also substantial damage to the area of the building.

This is a complex event. Possible contributing and root causes for this accident include:

- The employee used a "splash filling" technique to dispense the toluene rather than a safer drum pump.
- The static charge was not "drained away" to prevent discharge and ignition of the toluene vapors. A drum pump with a self-bonding hose would have prevented the accident.
- The employee was working alone. There was no co-worker to correct his procedure, immediately extinguish the fire, or lead the employee to the safety shower.
- The employee violated safety procedures and was not adequately trained in the hazards of dispensing flammable liquids.

There are also management system failures in this example. For instance, there is a problem here with implementing OSHA's Hazard Communication requirements (29 CFR 1910.1200):

- There was no initial or annual training.
- The laboratory did not keep an inventory of hazardous chemicals used.
- There was no written Hazard Communication program for the facility.
- The employee did not have access to a material safety data sheet describing the hazards and precautions for handling toluene.[2]

Note that this accident is highly complex and should be investigated by a team led by the area supervisor or manager. . The facility safety coordinator has an important role: helping to guide the team through the investigation process. A common mistake here is to expect the safety coordinator to conduct the investigation and write the report.

Major Process Related Accidents and Incidents

OSHA calls out minimum requirements for investigating covered process-related incidents in the PSM regulation (1910.119(m)). This performance-based standard is also an excellent guide for investigating personnel incidents of moderate to high complexity.

OSHA states "The employer shall investigate each incident which resulted in, or could reasonably have resulted in a catastrophic release of highly hazardous chemical in the workplace." The code goes on to require the investigation be initiated no later than 48 hours following the incident and that a multi-disciplined investigation team (including non exempt employees) do the work. A report must be completed and maintained for five years, providing a description of the incident, contributing factors and recommendations. A *system for addressing report findings and timely implementation/documentation of corrective action* is the key component of this standard. This systematic approach, with effective employee involvement is a critical feature of accident-incident investigation that builds trust with employees.

Readers are advised to clearly define the scope of accidents and incidents to be investigated. Certainly, highly visible events like fires, explosions and offsite toxic releases must be investigated using the OSHA PSM criteria. Events causing legal challenges and/or business interruption beyond a certain dollar figure should also be investigated by a team, very possibly led by an external expert.

Near Misses and Potential Hazards

The OSHA language "could reasonably have resulted in a catastrophic release" also requires employers to investigate near misses (a near-accident) and potential hazards. Consider the following near misses and potential hazards:

- Repeated operator failure to observe high tank level indicators and alarms
- Small, on-site releases of chlorine without personnel exposure
- Employees handling corrosive acid without eye-face and hand protection.
- Entry of a non-approved powered fork truck entry into a hazardous classified process area
- Incidents without severe consequences involving emergency shut-

downs and/or blown rupture disks, opened pressure relief valves, activation of sprinkler or deluge systems, and callout of the emergency squad

- Process deviations such as line pressure exceeding operational and safety limits
- A incident during confined space entry that partially engulfs or traps an employee
- Other equipment failures or situations that could have resulted in an accident or incident

The 1984 Bhopal, India Incident

This well-known and often studied toxic release incident involving release of highly toxic methyl isocynanate (MIC) emphasizes why actual case studies are so valuable in teaching people the principles of accident investigation.

The plant was partially owned by Union Carbide, and local concerns, producing pesticides. Plant siting (now an OSHA required safety assessment item) was not a concern when the plant was built. However, during the next 10 years, the population directly around the plant grew to the point that many civilians were directly at-risk.

The huge MIC vapor release (estimated to be 25 tons) killed over 2,000 civilians and injured an estimated 20,000 others. An interesting sidelight to this story is that no plant workers were harmed and no equipment damaged.

Plant management made two costly errors in judgment that led to the accident:[1]

1. There was a belief that growth of the shantytown did not pose a safety hazard.
2. A refrigeration system designed to maintain the MCI in liquid form had been shut down five months before the accident because it was believed unnecessary.

Analysis of this incident in hindsight proves instructional. There had been no initial or periodic PHA studies to identify or address a concern about location of the MIC equipment or other hazards of the process. Apparently the technical and safety staff did not fully rationalize the refrigeration safeguard, especially the consequences of losing or disabling the system. There was no policy for triggering management of change procedures in the event that process safety information is changed (technology of the process and protective safety systems).

The Bhopal accident is an example of a highly complex event. For this type of occurrence, Indian government investigators were involved in addition to facility and corporate personnel.

Staffing and Resources

In cases like the Bhopal release and the 1989 Phillips 66 explosion, the complexity requires a comprehensive set of skills available to an investigative team. This approach also applies to investigating accidents of a considerably more simple nature.

For example, understanding the behavior of flammable liquids and vapors and the interaction of controls and field devices requires people with technical backgrounds. Experienced process operators, process engineers, and maintenance personnel are needed to examine process and instrumentation diagrams (P&IDs), determine failure modes, and to review operating and maintenance procedures. Furthermore, investigators with interviewing skills are also needed to objectively gather the facts preceding the accident.

Assembling an investigation team takes some forethought and planning. Some facilities have more potential for complex accidents than others do. Chemical processing facilities generally require daily professional safety and industrial hygiene services. These skills are also necessary on most chemical investigative teams. Personnel that should be considered for accident investigation teams include the following:

- Operators and other hourly employees
- Product and process engineers
- Supervisors and/or team leaders
- Safety and health personnel
- Production managers and/or the facility manager
- Staff members such as industrial hygienists, health physics, corporate safety, environmental technology and services, etc.

Accident and Incident Investigation Policy

It is useful for a facility to develop a policy for investigation of accidents and incidents that meets company and other relevant standards. For example, Minnesota OSHA has a regulation called "A Workplace Accident and Injury Reduction" Act (AWAIR) that requires employers to write an accident prevention program. To meet the AWAIR law, the program must state how accidents will be investigated and corrective action implemented. Regardless of legal requirements, an aggressive accident/incident and potential hazard investigation program is good business.

Policies, procedures, and standards for performance need to be in place so that there is a quick and effective response to all accidents and incidents. There are a few key values that should define management expectations in the form of an accident investigation policy.

1. A pre-planned, immediate, and appropriate set of emergency response actions should be taken in the event of an accident or incident in order to eliminate or control the hazards responsible for the event.

2. The department head where the accident occurred should immediately launch a preliminary investigation to determine the immediate cause and corrective action plans. The individual then needs to communicate a preliminary alert (not to be confused with the formal accident investigation report) to all who need to know.

3. Appropriate documentation needs to be completed for all accidents and incidents such as the first report of injury form and forms meeting OSHA recordkeeping requirements (Form 300 and 301).

4. A formal investigation should be initiated for all accidents/incidents classified as serious or which have the potential for serious consequences. The immediate cause, causal factors, root cause, and corrective action to prevent recurrence need to be reported.

5. The department head and safety committee need to monitor corrective action activities until complete.

Clearly, line department heads are to lead the preliminary and formal investigation of all accidents occurring in their area. It is helpful to include safety and health personnel and other employees, especially for more complex cases. Departments are expected to share lessons-learned with other facilities operating similar processes. Investigation of significant process-related incidents should involve the appropriate technology manager.

Accident/Incident Investigation Procedures

All facilities need to develop an investigation process or procedure that applies to accidents, incidents, near accidents, and potential hazards. To avoid the common mistakes mentioned earlier in this chapter, it is critical for a facility to develop and install a clear investigation procedure like the one illustrated in Figure 4-1.

Eight Step Incident Investigation Procedures

It is important to understand and practice the basic steps necessary for an effective accident investigation ahead of time. Without a procedure in place, there is little chance that a scramble of unrelated activity after the accident will produce any meaningful change preventing future related accidents.

There are many methodologies (some proprietary, some free, and some at a price) that can be used for accident investigations. Whatever the methodology, plants need their own single procedure that can be applied to any accident, whether simple or complex. It is far easier to take un-needed steps out of an accident investigation procedure than to apply a "band-aid" to a poor or missing procedure. The following procedure applies to personnel and process-related accident investigations.

Step 1—Initiate Emergency Action Plan

When an injury-causing accident such as a chemical release, fire, or explosion occurs, the facility should be able to immediately execute the appropriate action plan. All employees of the plant should be trained to put the emergency action plan into motion. Review emergency planning and response guidelines in Chapter 2 and Figure 2-2 for the capabilities required of an emergency response team. In most accident situations, the co-workers and/or the area supervisor first respond to an injured worker. The next priority is to secure the accident scene. It is important to protect the boundaries of the accident scene from others not involved.

Most facilities have a central location like the security desk for forms and first-aid supplies. These supplies are often called the "Go Kit" (Figure 4-2).

The Go Kit carrying case should include an investigation procedures packet with paper, pencil, data, and interview forms; clipboard; flashlight; rolls of caution and duct tape; ruler and tape measure; digital camera; plastic bags and envelopes for samples; sealable containers with lids;

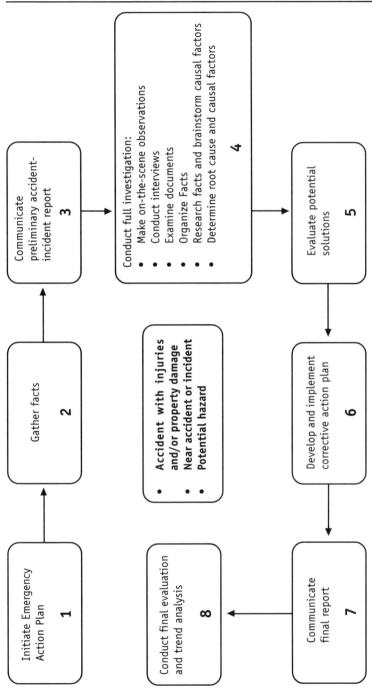

Figure 4-1 Accident-Incident Investigation Procedure

Figure 4-2 Go Kit[3]

blood-borne pathogen kit; rubber gloves, safety goggles, and other necessary PPE; and a clock.

Once an accident has occurred, supervisors should tell nearby employees:

* To leave the accident site but to stay nearby to help investigators as needed
* To avoid touching or moving equipment, tools, or work, as well as personal belongings

Step 2—Gather Initial Facts

Once the victims of an accident have received medical attention and the scene is secure, the investigative team begins to gather initial facts about the accident.

This is the time to photograph and/or videotape the scene to record information to be examined later. View the accident scene from several angles. Use a tape measure and record important distances. Having more information than needed is better than not having enough information.

Careful observations should be made of the work and related areas. Document and retain as much information as possible from the site. Record evidence such as incident date and other information related to when and where the accident occurred. It may be helpful to use a checklist or form that includes only the basic facts of the accident and no in-depth analysis (Figure 4-3).

Step 3—Communicate Initial Accident-Incident Report

The initial accident-incident report is the notification that an accident or incident has occurred. This initial report fulfills the early reporting requirements to divisional and corporate groups as well as to facilities with similar processes. Figure 4-3 provides the format for a typical accident/incident report.

Note that the initial alert supplements reporting requirements for regulatory agencies such as OSHA and EPA. For example, OSHA recordkeeping requirements now include form 301 (Injury and Illness Incident Report). If injuries are involved, this is also the time to complete and attach a *First Report of Injury* form for worker's compensation and insurance purposes. Your insurance carrier may issue these forms (or may accept a facility initial accident report form) or you can use OSHA 301 form.

We recommend this step be called an *initial report, incident alert,* or something similar to distinguish it from a final report. At this early stage, you do not want to give employees and others the impression that the investigation is in any way complete.

Step 4—Conduct Full Investigation

When setting up an accident investigation procedure, make sure to distinguish between simple, personnel-type accidents and complex, process-personnel accidents and incidents. For personnel accidents of low complexity, the National Safety Council recommends the investigation be done by the immediate supervisor of the involved employee(s), or in some cases the department head.

In more complex situations, it is extremely important for management to appoint an investigation team (including consideration of outside experts). In fact, OSHA's Process Safety Management regulation requires that a multi-disciplinary team initiate the investigation no later than 48 hours following the incident.

Remember that the makeup of the investigation team is part of your advance planning. Consider all personnel and functions that can provide skills appropriate to the situation. While selecting the team, keep in mind the various aspects involved in the accident, such as the manufacturing process, safety and health problems, and the management structure of your organization. The team chairperson should be someone who was not directly involved with the incident and should be selected based on his or her ability to remain objective throughout the investigation.

Incident Date: _____   Incident Time (Local) _____
Site Location: _____   Division: _____

Department/Area of Incident: _____

Name(s) of injured: _____

Description of injuries: _____
_____

Description of event:_____
_____
_____

Reported by: _____   Phone: _____

Follow-up contact: _____   Phone: _____

Type of event: _____ (see code below) Component: _____ (see code below)

Equipment or process involved:
_____

Location within facility: _____

Description of property damage: _____
_____

Approximate damage cost: _____

Describe emergency response:_____

Preliminary cause(s) known or probable: _____
_____

Immediate plans: _____
_____

Assistance needed: _____

Person receiving report: _____ Date: _____ Time: ____

OSHA/EPA/DOT notification required:_____

Incident Type: F Fire   E Explosion   HR Hazardous release
               PR Pressure release V Vacuum loss   0 Other (specify)

Component: T Tanks   P Piping   RC/TR R. Car/Trucks
               PMP Pumps   VL Valves   H Hoses
               FL Filters   0 Others (specify) Ch Chemical agent

Copy list:

Plant/General Manager_____

Safety/Health/Environmental Manager/Coordinator_____

Department Head_____

Other (list) _____

Figure 4-3 Model Initial Accident/Incident Report

Below are examples of additional participants to consider for the team depending on the type of accident or incident investigation.

Type of Accident/Incident	Additional Participants to Invite
• Actual or potential event involving a spill or release	Environmental protection department
• Accident resulting in actual or potential:	
-fatality, recordable injury, and/or illness	Facility manager, outside expert, Plant/corporate safety, legal
-greater than $5,000 loss	
-Serious incidents without injury.	
• Any of the above in which maintenance contractors were involved	Contract services group
• Any of the above in which construction was involved	Construction group

Only after a team has been created can a full accident investigation, including the following six aspects, take place:

a)   Make on-the-scene observations

b)   Conduct interviews

c)   Examine documents

d)   Organize facts

e)   Research facts and brainstorm causal factors

f)   Determine root cause and causal factors

Make On-The-Scene Observations

Investigators must keep in mind several questions when reviewing on-the-scene observations:

• What was the availability of help or mechanical aids to the work being done?

• What if any lockout/tagout or line-breaking devices were in place at the time of the accident?

• What PPE/clothing was the injured person using when the accident occurred? What was the condition of PPE/clothing after the accident?

• What was the condition of critical safety devices/systems following the accident (valve position, high/low level controls/alarms, rupture disks, etc.)?

• What was the condition of the worksite related to accident avoidance and self-rescue (clear access, blocked access/aisles, materials or work outside of normal work area)?

Drawings or diagrams are particularly helpful if they are drawn to scale and include measurements of equipment involved and the relationship between objects. Only the facts related to the immediate cause are of concern. Again, see Figure 4-3 to prompt information needed.

Record fragile or perishable evidence and information, including instrument and control panel settings/readings, work environment conditions, and weather information. Photographs or videotapes of the accident scene can be invaluable when analyzing causal factors. Make sure that as much of the accident scene as possible is recorded from a variety of angles. Having more evidence than needed is better than not having enough evidence.

Conduct Interviews

Witnesses to the incident, including the employee(s) directly involved, are important sources of information and should be interviewed *as soon as possible* following the event. By promptly conducting the interviews, we avoid the problem of witnesses getting together to reach some kind of consensus regarding what was observed during the incident. Witnesses (and all employees) should be asked not to discuss the case with each other until after all have been interviewed; however, this is difficult to enforce.

The witness should be interviewed one-on-one *at the incident scene,* if at all possible. Surroundings and the behavior of the person conducting the interview make an important difference on how the witness may respond. The last thing you want to do is to talk down to the witness with a desk between the two of you. Because witnesses will often be nervous or fearful, you want to establish a climate of calm and trust immediately. It is important to say that you are simply *looking for facts, not trying to place blame.* Do not take a lot of notes; this can be intimidating to a witness. It is better to quickly jot down critical information and fill in the outline of the conversation immediately after the interview.

When appropriate, ask the witness to physically demonstrate positions of people at the time of the incident. Use a consistent outline of questions for all witnesses. Ask questions such as:

- Who was injured and what is your relationship to the injured person(s) (co-worker, supervisor, etc.)?
- Who else was in the area?
- Who was operating the equipment/process?

- Who installed or repaired the equipment?
- What happened?
- What did it sound like (relate it to something similar in sound and volume)?
- What did you feel (relate it to something similar in size and motion)?
- What was the injured person doing at the time of the accident?
- What were you and others doing?
- What tools, machines, materials, and chemicals were involved?
- What were the conditions of the workplace?
- What potential hazards were identified prior to the task involved and what (if any) measures were taken to prevent the incident?
- Where was the worker standing (show me)?
- Where was the fork truck, ladder (or equipment and materials involved)?

You want a minute-by-minute account of events preceding the accident, so keep questions detailed and specific. For example:

- When did the accident occur?
- When did the rupture disk blow?
- When did you notice the smoke/flames, smell, or noise?
- When did you/he/she activate the vent control?
- How did he/she get sprayed with the acid?
- How did the chemical/water get on the floor?
- How long has the job been done this way?
- How are written procedures (have this in hand) followed?

Avoid subjective "why" questions. Instead, keep "why" questions specific and objective, such as:

- Why were certain equipment, chemicals, and procedures used?
- Why was the floor wet?
- Why did these events occur in this order?

Ask checking questions as follow-up to get more specific and factual answers.

The use of *open-ended questions* is recommended. Be a good listener. Let the witness tell as much of the incident story as possible without frequent interruption. However, it is important any long answers to open-ended questions be paraphrased. Make this exchange as conversational as you can. It is better to say "In other words, you (saw, heard, felt)"—than "Do you mean —?" when paraphrasing. Allow the witness time to correct your interpretation of what they said, or expand on their statements.

Document the information provided, and allow the witness to review your notes. Finally, sign and date the statement.

Examine Documents

The most reliable source of facts an investigation team has are records and other written information relating to the incident. If there is any doubt, compare the statement of a witness who says, "I think the machine had a problem," with inspection reports and maintenance logs that either do or do not document attention to the problem.

Look for documentation items such as:

- process instrumentation and control documents
- operating procedures
- training logs and manuals
- batch sheets
- finished goods and raw material quality control records
- run histories
- emergency responder logs
- product and material samples
- material safety data sheets
- past accident/incident reports
- records of process control software
- shift logs
- printouts of field instrument locations and valve positions
- printouts of electrical equipment and positions of switch devices
- repair and preventive maintenance logs

Facts from the document review will support your conclusions and recommendations. Make sure to identify and reference all relevant documents in the incident report. You are now ready to consider causal factors for the accident. The following is a simple problem-solving process that can be applied to almost any accident or incident.

Organize the Facts

Organize the facts relevant to the accident according to the categories of People, Machine, Method (Procedure or Behavior), Material, and Environment.

Example 1

Consider splash filling a static generating powder into a vessel partially filled with a flammable solvent (see Figure 2-1). Write the items on a fishbone diagram like the one pictured below (Figure 4-4).

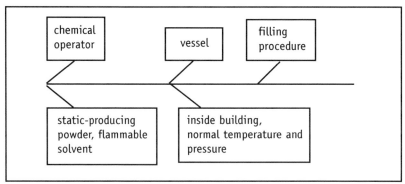

Figure 4-4 "Fishbone" Diagram

Once organized, ask "What/Where/When is the problem? Who is involved? Who is affected?" Define the problem by describing what you know (do not assume) actually happened versus what should have happened if the scenario was problem free. Is the problem in one specific area of the facility, in several areas, or does it exist everywhere? Does the problem occur only under certain circumstances (for example on second shift or during shift changes)? Are some employees affected by this problem more than others (e.g. process operators, maintenance personnel, packaging employees) are?

Research Facts and Brainstorm Causal Factors

This stage of the problem-solving process involves carefully researching each fact discovered and brainstorming possible casual factors contributing to the problem. Many of what seem like obvious causes will be wrong; therefore, the team must keep asking why until there are no more responses. It is also important not to make judgments while brainstorming since it tends to choke off the flow of ideas.

In the example above, we may consider a casual factor to be that the safety auger was not used to feed the powder into the reactor. Another factor, further upstream from the accident, is that the operator was not trained to use the safety auger.

Example 2

Another helpful way to determine the root cause and causal factors is to write each event that contributed to the accident on *Post-it*® Notes and put them in a time sequence. Again, in our splash-filling example, this might look like the following figure (Figure 4-5).

Use a dashed rather than solid line if the event is inferred to have happened and is supported by the facts.

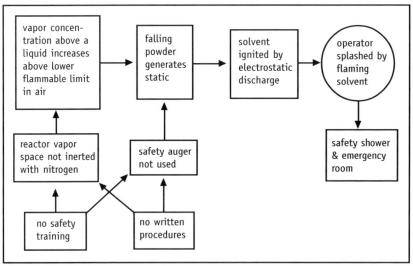

Figure 4-5 Root Cause Analysis

Example 3

Another approach to determining the root cause and causal factors is to create and work with a *category checklist* to prompt possible casual factors.

In a Battelle study of petroleum refining facilities, the authors propose the three categories of facility design, procedural factors, and management systems to account for examples of incidents involving human factors.[4] Nearly half of the accidents "involved elements of human error." Factors contributing to causes involving human error were measured to be:

Procedural	42%
Management systems	32%
Facility design	7%
Random human error	19%
Equipment failure	(no human error)

The Battelle authors state that 90% of the procedural causal factors reported had to do with failure of operations and maintenance workers to follow safe work practices. This finding is certainly no surprise in light of industry-wide statistics that demonstrate accidents are most often caused by unsafe acts rather than unsafe conditions. If written procedures exist, they are generally accurate although not always descriptive in what people are to do. A common problem in procedure writing is using language that is too technical for operators to thoroughly understand. Just as serious are tendencies for written procedures to be vague, inconsistent, and out of date.

An ongoing problem throughout industry is failure to follow written procedures. More serious however, is compulsive or habitual unsafe acts that lead to accidents. See the discussion on behavioral methods in Chapter 6. The Battelle study traces cause of 85% of the management system failures to problems related to training. Too often, required safety training for topics like lockout-tagout, confined space entry, and even process safety management is treated in a stand-alone fashion. That is, the safety training is not job-specific or integrated into procedural skills training. For specific ways to integrate safety and job skills training, see Chapter 9.

Finally the Battelle authors found that nearly half of the design causal factors are related to process controls as opposed to the equipment itself or the work environment. Here specific factors include controls that are not labeled, those that might be confusing because they are associated with many manual steps, and lack of controls altogether. Accident investigators frequently recommend corrective action where process and procedural understanding are the basis for control specifications.

Determine Root Cause and Causal Factors

After writing down all possible causal factors, the team is ready to identify those that are supported by the facts and those that are probable.

In the above example, we chose the electrostatic ignition as the immediate cause of the accident. It is the root cause as well. Failing to inert the vessel, the missing safety auger, lack of written procedures, and lack of training are the causal factors for this accident.

Step 5—Evaluate Potential Solutions

Once the investigation is completed, potential solutions must be evaluated. This step involves brainstorming and evaluating the corrective action measures that will eliminate or mitigate the accident causes determined in step four. The key to this step is to concentrate on prevention and avoid treatment of symptoms. Evaluate each potential solution against the *Safety Precedence Sequence* (SPS) found in Figure 4-6.

It is suitable to have a number of solutions meet the requirements of the SPS. Select the solutions to be implemented. Whenever possible, adopt engineering controls that eliminate the hazard (cause). If that cannot be done, choose solutions that meet priorities 2 or 3 of Figure 4-6. Make sure management has an opportunity to be part of decisions that require unbudgeted resources.

Propose solutions that will address the most probable causes of the accident. Provide a general plan for each solution, including what is to be done, who will be involved, and resources required. In addition, evaluate each proposed solution according to the following criteria:

- There is likelihood that the problem will be corrected.
- No additional serious problems will be created.
- The cost versus benefit is justified.

(In order of decreasing effectiveness)

1. Use engineering controls. Design the solution to reduce or eliminate the hazard.

2. Install automatic safety devices like pressure relief valves and rupture disks and/or fail-safe process control interlocks.

3. Install automatic safety warnings (alarms, sirens, console lights, permanent signs, etc.)

4. Develop or improve written procedures including specified personal protective equipment (PPE).

5. Take some personnel actions (e.g. increased supervision, classroom or on-the-job training, adjusted work hours, and disciplinary action).

6. Identify residual hazards and risks to the appropriate level of management.

Figure 4-6 Safety Precedence Sequence

Step 6—Develop and Implement a Corrective Action Plan

A corrective action plan really begins with damage control and restoration of normal operations. Therefore, this step begins well in advance of the full investigation. One must act cautiously in restoring the process, so that valuable evidence is not destroyed. Jumping to conclusions as to the root cause of an accident is another common mistake experienced during early corrective action activities.

Example 1

In a low-complexity accident, concentrate on eliminating the immediate cause. In the example at the beginning of this chapter, the worker was required to wear PPE and demonstrate that he or she knows how to correctly add acid to water.

Example 2

In higher-complexity cases, the root causes and causal factors frequently involve the safety management system and take more time to correct. In the toluene fire example, there were several parts to the corrective action plan:

1. Required use of a metal receiver, drum pump, and self-bonding hose for transfer of flammable liquids.
2. Formation of an in-plant emergency squad and instruction in fire extinguisher use for all employees.
3. Development and implementation of a flame-retardant clothing program.
4. Use of Job Safety Analysis and Job Instruction Training to improve compliance with written procedures.
5. Implementation of a written Hazard Communications program conforming to company policy and OSHA requirements.

The corrective action plan defines what will be done, by whom, and when. Describe in detail how the solution is to be implemented (e.g. initiate shop work order to inspect and service pump, initiate new procedures, and begin new operator training); establish priorities according to level 1 (most important), level 2, or 3; and estimate start and finish dates (30-, 60-, or 90-day intervals). Assign responsibility for each corrective action to be taken. Make reference to the Safety Precedence Sequence to indicate the effectiveness of the solution. Where appropriate, schedule a pre-start review meeting to see that all action items are complete before restarting equipment (required by the OSHA Process Safety Management standard).

Avoid references to the injured employee, location, or date of the accident. Use discretion that will avoid unwanted emotional reactions that may compromise the solution.

Example 3

For the splash filling accident above, action items 3, 4, and 5 from example 2 apply. Note that OSHA 1910.106 and .108 plus NFPA 30 all name specific requirements for storing and handling flammable liquids that can be adapted to your facility's corrective action plan.

Step 7—Communicate Final Report

Only when the team has analyzed all the facts and determined the root/contributing causes are they ready to begin compiling a report. The critical skill of this step is to document corrective actions that address each factor involved in the accident and to make sure each recommendation is supported by facts and conclusions.

We have seen that the initial accident report announces the investigation; the final report closes the case. Again, the importance of avoiding the rush to judgment by circulating an incorrect or inconclusive report prematurely cannot be overemphasized. Figure 4-7 represents an outline that can be used to organize a final report.

(Remind your accident investigation team that Figure 4-7 is not to be used as a *form* to fill out without going through all the stages outlined under Step 4.)

Avoiding Common Mistakes in Report Writing

The final challenge in an effective accident investigation is to put the results together in a meaningful way. Unfortunately, there are many mistakes made in report writing.

Recommendations should address each root cause. Investigation reports that are incomplete are not fully objective. In addition, each recommendation needs to be supported by the facts and conclusions for the accident.

To avoid stating information incorrectly, the writer should have the team screen a first draft. It is necessary to have the right choice of words describing facts, conclusions, and recommendations.

TITLE

SUMMARY

Facts from Initial Report (Figure 5-3)

_____

_____

_____

_____

Analysis and Conclusions

_____

_____

_____

A. Causal Factors

_____

_____

_____

_____

_____

B. Root Cause

_____

_____

_____

Recommendations

(Corrective Action Plan)

_____

_____

_____

_____

Figure 4-7 Accident Investigation Final Report Format

Attention to writing style is important. Make sure there are transitions and continuity from one section of the report to another. Avoid inflammatory language and editorializing.

The purpose of the report should be clear from the beginning as to its purpose. Accidents are investigated to identify root causes in order to avoid similar accidents in the future. Thus it is critical to screen the report in order to remove any words that imply opinions, values, or judgments. Examples of words that may raise a red flag include the following (and their negatives):

Accessible	Complete	Assured
Acceptable	Effective	Expected
Adequate	Approved	Functional
Allowable	Available	Needed

Examples of words that imply conclusions are (and their negatives):

Affected	Indicated	Aware	Knew
As a result of	Leading to	Capable	Sure
Because	Hampered	Experienced	Skilled
Due to	Should	Intended	Trained
In order to	Able	Intentional	

Finally, examples of words that contain implied values include (and their opposites):

Average	Easy	High	Normal
Clean	Faulty	Hot	Poor
Clear	Fair	Large	Mediocre
Difficult	Good	Limited	

If the types of words above survive the screening process, make sure that, in context, they say exactly what you mean.

Developing Facts, Conclusions, and Recommendations

Facts are statements that need to be objectively verified. Recommendations are calls for action. Coming to the right conclusion about the basic or root cause of the accident, based on the facts, is critical.

Conclusions almost always involve evaluations, assumptions, judgments, and cause-effect statements. Thus the investigation team needs to test the premise (presumed accident scenario) to make sure it holds up under discussion. A good rule of thumb is to make sure a conclusion is supported by at least two facts.

Recommendations or recommended corrective actions are the team's call to action by management. These actions need to directly address *all causes*. Go back to the Safety Precedence Sequence, Figure 4-6. The categories may be reduced to the sequence of engineering, procedural, policy, and training changes. Corrective actions are based on the conclusions about the accident. Recommended actions are the most effective way to solve the problems that led to the accident and should include cost/benefit comparisons as well as identification of the personnel that will implement the solution. Do not make recommendations too restrictive. Provide enough guidance to define the action, but do not write the recommendation so specifically as to restrict the corrective action plan. It is important for the team to develop alternative recommendations in case one or more of the corrective actions is challenged. The action items need to be prioritized into short and long-term tasks.

Finally, the team should document the assignment of action items and describe any residual risk remaining.

Step 8—Conduct Final Evaluation and Trend Analysis

This is the final step of the entire procedure, the purpose of which is to assure that the goal of preventing future accidents has been met. Once corrective action items have been completed, the original investigation team or the safety committee should revisit the scene to see that the problems addressed have been eliminated or reduced.

During this visit, at least three questions should be asked:

1. Were the recommended action items effective in dealing with the problem(s)?
2. Were other significant problems created by the corrective actions?
3. Is there more to be done to prevent accidents of this type?

Incident and Causal Factors Database

Documents and records pertaining to incidents are the most reliable source of facts that support conclusions and recommendations developed from the investigation. In a similar fashion, information gathered from incidents of various types has long-term value to a facility. For example, performance changes over time can be tracked if the information can be accessed in a meaningful way. Perhaps the most important application for an incident database is to provide statistical data for process safety assessments and audits. Another use for a database is to focus special attention on hazards that have been the cause of previous incidents. Use the following criteria to set up an incident database:

1. It must be easy to use. Insertion and retrieval of data must be simple and consistent for those who maintain the database.
2. The information within the database must be protected from possible tampering. Although the information should be widely available, only authorized personnel should be able to change or copy it.
3. New or revised information must be added promptly so that the database is kept up to date.

Progressive organizations have databases to build, document, and maintain hazard and risk information. Commercial software is readily available that helps organizations set up and maintain a database for storage of investigation results and corrective action data.[5]

In summary, a good accident investigation identifies and controls hazards that often have been long neglected. Effective accident investigation requires planning, and planning and following an accident-incident investigation procedure is critical to success.

Potential Process Incident Investigation Procedures

OSHA Process Safety Management Requirements[6]

OSHA requires a team approach for process-related accident investigations. The requirements include:

1. Investigation of each incident that resulted in, or could have resulted in, a catastrophic release of toxic or flammable materials
2. Prompt initiation of investigation procedures, no later than 48 hours following the incident
3. Formation of a multi-disciplinary team including at least one person knowledgeable in the process involved and a contractor employee if a contractor was involved
4. Preparation of a report that includes relevant dates and a description of the incident, casual factors, and recommendations

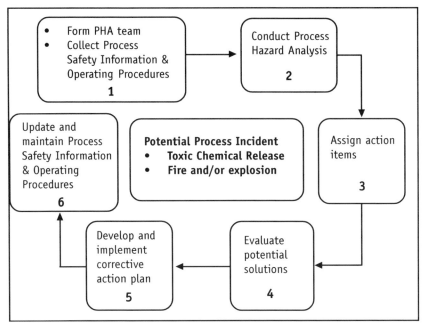

Figure 4-8 Investigation Sequence for Potential Process Incidents

5. A management system to promptly address and resolve report findings and recommendations
6. Review of the report with all affected personnel whose job tasks are relevant to the report findings
7. Retention of incident investigation reports for five years

Accidents and incidents that are major require special notification, investigation, and reporting procedures for the following reasons:

*   Extent and complexity of the damage
*   Technology involved
*   Regulatory considerations
*   Potential for large dollar losses and claims

The main tool from OSHA's Process Safety Management regulation that can be used for incident investigation is the Process Hazards Analysis (Step 2 in Figure 4-8).

Process hazards are unidentified scenarios or incidents waiting to happen. Thus a PHA methodology functions as a protocol for investigation of process-related incidents, and the PHA team becomes a potential incident investigation team as they consider the following:

- Hazards resulting from deviations in procedures or from the process itself
- Previous incidents that had the potential for catastrophic consequences
- Interrelationships between engineering and administrative controls
- Consequences of failure of engineering and administrative safeguards
- Location of rooms and buildings relative to sources of fire, explosion, or toxic release
- Potential for human error
- A qualitative evaluation of possible safety and health effects of failure of other safeguards for employees, the public, and the environment

The "how-to" details for conducting PHAs are provided in Chapter 5.

Once the PHA study is completed, the remaining steps in Figure 4-8 are almost identical to those followed during a personnel-type accident-incident investigation procedure (Figure 4-1).

Incident Case Studies

Example 1—Bhopal MCI Release

Had a process hazards analysis (PHA) been conducted, the Bhopal incident described earlier in this chapter may have been prevented. Causal factors for this event involved lack of attention to facility siting and a management of change policy. The reader will recall that location of the plant relative to the community shantytown was initially not a problem, since the plant was located more than a mile from a populated area. Of course, the OSHA PSM standard was not in effect in 1984 plus OSHA requirements do not apply to facilities outside the U.S.

This case reinforces the value of facilities using initial and revalidation PHAs as management policy, not simply because it's the law. The PHA tool is effective not only as an accident investigation tool, but primarily in preventing catastrophic incidents like Bhopal. The process of identifying process deviations, previous related incidents at the current or other site, engineering and administrative controls plus their interrelationships, facility and equipment siting and human factors facilitates discovery of dangerous but plausible scenarios. Revalidating the PHA (required by OSHA every 5 years) would have prompted the study team to re-examine changes that occurred since the plant was built 10 years before the release.

Use of a functional management of change policy is another way this accident could have been prevented. When the idea came up to deactivate the refrigeration system, the policy would have triggered a review of the technical basis for the proposed change, its impact on safety and health, necessary modifications to operating procedures, time period required for the change, and most importantly, the authorization requirements for the proposed change. A qualitative or quantitative risk analysis was likely to have preceded authorization to decommission the refrigeration system.

Example 2—Vulnerability of Control Rooms

Work done at ICI Chemicals and Polymers in the U.K. applied to accident investigation results from the disastrous Flixborough England explosion in 1974 and to other situations where control rooms are vulnerable to fires and explosions.[7] In the Flixborough accident, 28 people were killed in the Nypro Chemical Works control room.[8]

In this case and in other situations where ICI reduced the vulnerability of control rooms, risk analysis was used to justify what is usually a high cost solution. Top managers make business risk decisions all the time. Risk is the product of the consequences of an unwanted event times the probability that it will happen. Moving the control room eliminates the hazard and the study team does not have to debate the event's likelihood.

Given availability of tools like PHA and the important results of studies like the one by ICI it seems examining and reducing risk to people outside, in buildings and rooms near chemical processing operations is an issue that receives too little attention. Also, advancement of networked distributed control systems (DCS) technology makes unnecessary the placement of control rooms in the immediate vicinity of hazardous process operations.

Example 3—Hazards Resulting From Air Emission Controls

In an effort to eliminate or reduce air emissions, more and more organizations are installing solvent recovery units, thermal oxidation systems, carbon canister filters to absorb volatile organic compounds (VOCs) and other control systems. Unfortunately, this environmental protection equipment may introduce new hazards into the system like fire, explosion and/or employee toxic exposure.

This example begins with results of a survey (1990-1997) of the member companies of the American Petroleum Institute's (API) Safety and Fire Protection Subcommittee (SFPS).[9] The survey determined the number and type of air-emission controls installed and the kinds of incidents which participants have experienced.

Responses were received from eight SFPS companies that operated eight refineries, one gas plant, two chemical plants, and one marketing terminal.

A total of 2,110 air-emission controls have been installed at these facilities, and 70 incidents associated with these controls have been reported. The incidents were split roughly half and half between recurring events associated with the same control and so-called "unique incidents."

The following is a summary of the air-emission control device hazards identified during investigation of the incidents:

Hazard	Responses
• Employee exposure	58
• Flammable atmosphere due to enclosure	43
• Flashback of flammable vapors	23
• Overheating of carbon canisters	20
• Loss of flame	20
• Overpressure	13
• Other	6

Carbon canisters are the most commonly used device used in refinery sewer systems to absorb VOCs. In the study, there were 10 unique (five events resulting in injury or major property damage) and 10 reoccurring carbon canister problems.

The author is aware of another incident where a carbon bed, used for solvent recovery, became saturated with the VOC. The solvent vapor escaping from the bed quickly reached its lower flammable limit. The mixture exploded when it came to the hot face of a downstream thermal oxidizer. Miraculously, there were no injuries; however significant property damage occurred.

These cases reinforce the importance of investigating near misses and actual and potential accident scenarios in order to prevent future incidents similar in nature. Accident investigation and hazards analysis lead to process understanding. Process understanding leads to improved procedures, controls and operability. From a business and safety standpoint, this is a win-win situation.

Example 4—Time Sequence Events Before the Accident

This valuable lesson comes from the disastrous Hindustan Petroleum Corporation (HPCL) refinery incident in India (1997).[10] An undiscovered leak, caused by corrosion of a liquid petroleum gas (LPG) vessel at the refinery, generated a vapor cloud, which led to two explosions and fires lasting for three days. Sixty people were killed and there was over $15 million in property damage, including destruction of 19 buildings and 25 storage tanks. Estimates made at the time indicated it would take a full year to put the refinery back in operation.

The authors provide a time sequence of the accident (Figure 4-9).

The investigation showed that the first vessel, located near the main gate and filled with LPG/crude/kerosene, began to leak, quickly caught fire and exploded at 6:40 a.m. A second vessel blew up 15 minutes later and before noon, fires blocking both entrances to the plant produced intense flames and thick black smoke. A quick rain shower brought soot down on the heads and shoulders of people nearby and the runoff flooded the roads with murky water.

The facility had no fire-fighting equipment and Navy emergency responders could not gain entrance into the facility because of the fires. It took 3 days to consume the burning fuel.

A look at the time sequence of recorded events indicated that the original leak was not discovered for 12 hours and that corrective action was not taken before the first explosion. The article's authors point to a "lack of attendant/supervisor responsibility and communication plus a general lack of safety consciousness." However, there are basic management systems problems at work in this accident. Had a policy been in place to supervise all unloading and loading of flammables, the leak would have been discovered right away. In addition to the lack of communication, there was no system to communicate and activate the need for a rapid response to prevent the flammable vapor from reaching an ignition source. The authors point out the lack of an emergency plan; in addition, there was no fire safety infrastructure to provide fire-fighting equipment and training for employees. There was no possibility for mutual aid from

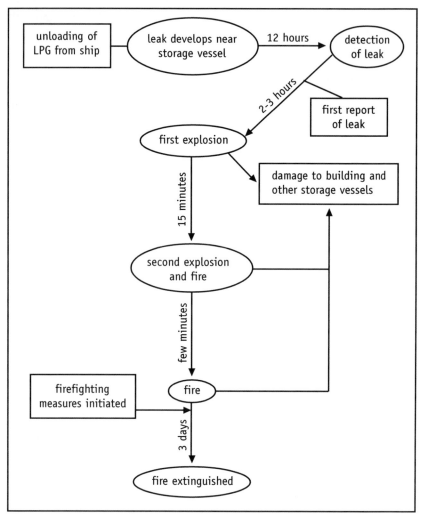

Figure 4-9 A Sequence of Steps Involved in HPCL Accident[11]

outside emergency responders because the facility did not have an emergency squad to form a relationship with these people.

Lack of a system to conduct hazard studies was another facility management gap. A "What-If" analysis may well have revealed the need to provide alternate access to the plant or move flammable liquid tanks stored near the two entrances.

Note On Legal Involvement and Corrective Action

The reader should consider the potential effect of legal approval of accident investigation reports. You certainly do not want censorship to result in a document that does not reflect the report submitted by the accident investigation team or that compromises corrective action taken by the plant. For example, there may be a call to remove recommendations such as "Provide additional PPE training" or "Change to a more comfortable goggle style." Make sure the legal department is fully involved from the beginning of the investigation and that sensitive findings are protected by attorney-client privilege.

Conclusion

It is easy but very damaging to simply blame the victim of an accident for "worker error." The purpose of accident investigation is to prevent reoccurrence. Therefore, every accident, incident, near-miss, and potential hazard should be reported immediately and investigated at a level appropriate to the actual or potential consequences.

Initial reports, or accident alerts, need to be concerned with immediate plans only. Investigation procedures involve thorough fact finding and careful analysis of causal factors preceding the incident. The critical part of the final report is a set of recommendations and a corrective action plan that is supported by the facts.

Facility management needs to develop an accident investigation policy and make sure related procedures follow a clear sequence or process. Documentation must include provisions for simple retrieval of information that can be used for hazard identification and control activities. Recommendations are to follow the safety precedence sequence (SPS), "engineering-out" causal factors and other hazards whenever possible.

It is a mistake to expect the safety coordinator to conduct accident investigations. To place ownership where it belongs, supervisors/managers and lead employees should be trained in basic accident investigation principles. Employees have an important role too. They are often assigned to be a part of the accident investigation team. If their workstation is near or part of the accident site, they may be involved in emergency rescue. Affected employees are often asked to leave the scene, but to remain nearby to assist in determining answers to questions involving where, when, how, and why the accident occurred. Finally, they should not touch, move, or remove work, materials, or equipment from the scene, even if the item includes personal belongings.

Documentation and analysis of previous and potential incidents plus external case histories also play an important part in accident prevention. The hazard study may be conducted before (process hazards analysis) or after (incident investigation) a potential or actual event.

References

1. Ralph King and Ron Hirst, *King's Safety in the Process Industries, 2nd edition*, Wuerz Publishing, Winnipeg, Canada, (co-publisher).

2. Toluene Material Safety Data Sheet, Ashland Chemical Co. (MSDS No. 301.0000565-010.008I, or current version), 6/21/99.

3. Anthony V. Ventresco and Larry R. Russell, "Team-Based Incident Investigation," *Professional Safety*, American Society of Safety Engineers, May, 1998. Reproduced with permission.

4. G. Bradley Chadwell, Fred L. Leverenz, Jr., and Susan E. Rose, "Contributions of Human Factors to Incidents in the Petroleum Refining Industry," *Process Safety Progress*, Vol. 18, No.4, Winter 1999, Table 1.

5. www.taproot.com. TapRoot is a registered trademark.

6. Code of Federal Regulations, 29 CFR 1910.119(m) *Incident Investigation*.

7. R.A. McConnell, "Risk Management Plans for Existing Control Rooms," *Process Safety Progress*, Vol. 18, No. 4, Winter, 1999.

8. UK Department of Employment, *The Flixborough Disaster Report of the Court of Inquiry*, HMSO, ISBN 011 3610750. (1975).

9. Henry Ozog, Arthur D. Little, Inc., Cambridge, MA. and William J. Erny, American Petroleum Institute, "Safety Hazards Associated with Air-Emission Controls," *Process Safety Progress*, Vol. 18, No. 9, Spring, 2000.

10. Faisal I. Khan and S.A. Abbasi, "The World's Worst Industrial Accident of the 1990's," *Process Safety Progress*, Vol. 18, No.3, Fall, 1999.

11. Faisal I. Khan and S.A. Abbasi, "The World's Worst Industrial Accident of the 1990's," *Process Safety Progress*, Vol. 18, No.3, Figure 2. Reproduced with permission of the American Institute of Chemical Engineers. Copyright 1999 AIChE. All rights reserved.

Safe Work Conditions

Introduction

Exposure to hazards causes accidents. Hazards are usually classified as unsafe work conditions, unsafe work practices, or most often a combination of both. The next two chapters address these two classifications and the means to prevent and control unsafe work conditions and practices.

Particular chemical hazards of interest are those that may cause toxic exposure to employees, chemical exposures, spills and releases affecting the community and environment, as well as fires and explosions.

Table 5-1 Major Events and Associated Hazards [1]

Main Cause	Percentage
Toxic Exposures	
Chemical reaction (Bhopal, India)	—
Overpressure (Seveso, Italy)	—
Large Fires	
Flammable liquid or gas release or overflow	17.8
Overheating, hot surface, etc.	15.6
Pipe or fitting failure	11.1
Electrical breakdown	11.1
Cutting and welding	11.1
Arson	4.6
Others	28.7
Explosions	
Chemical reaction	35.0
Combustion explosion in equipment	13.3
Unconfined vapor cloud	10.0
Overpressure	8.3
Decomposition	5.0
Combustion sparks	5.0
Pressure vessel failure	3.3
Improper operation	3.3
Others	16.8

(continued on following page)

Table 5-1 Major Events and Associated Hazards [1] (continued)

Explosions- Frequent Location of Occurrence	
Enclosed process or manufacturing building	46.7
Outdoor structures	31.7
Yards	6.7
Tank farms	3.3
Boiler houses	3.3
Others	8.3
Explosions- Various Contributing Factors	
Rupture of equipment	26.7
Human element	18.3
Improper procedure	18.3
Faulty design	11.7
Vapor-laden atmosphere	11.7
Congestion	11.7
Flammable liquids	8.3
Long replacement time	6.7
Inadequate combustion controls	5.0
Inadequate explosion relief	5.0

Table 5-1 is a representative list of major events and associated hazards compiled for the chemical process and refining industry.

The events listed in Table 5-1 are ranked from most common to least common. With the exception of arson, human element, and improper procedure, these hazards, to a large extent, involve unsafe work conditions. Thus, this list of incidents demonstrates the importance of identifying and correcting unsafe work conditions during physical inspections and studies of process hazards.

Personnel and Process Safety Program

Personnel and process safety programs are quite different when it comes to the visibility of the hazards and the means for controlling them. Controlling many of the hazards listed above is considered part of a personnel safety program. Most facilities have a program that includes some kind of workplace inspection system to identify and correct this type of unsafe work condition.

Process safety management (PSM), first proposed by OSHA in 1990, was intended to improve safe work conditions and practices in the petroleum and chemical process industry. Facility process safety programs are relatively new in comparison with more traditional personnel safety programs.

Identifying and controlling process hazards, which are frequently less visible than personnel hazards, requires a different approach than direct inspection. Facilities need to use process hazards analysis (PHA), a key element of OSHA's process safety management regulation.

Focus physical inspections on actual or potential hazards that cause worker exposure. Focus hazard studies related to PSM primarily on potential process system upsets or a combination of equipment/human failures that could cause a catastrophic event.

Management System

Safety and health in the facility is as manageable as anything else done at the facility, and assurance of safe work conditions is a key part of the chemical safety infrastructure. A strong management system is needed for workplace inspections and identification/control of process hazards—one that provides effective corrective action once the hazards have been identified. The tools presented in this chapter will greatly aid any facility in meeting these goals.

Physical Inspections and Followup

The purpose in conducting these inspections is to locate, report, and initiate corrective action for existing and potential hazards so they can be corrected before an accident occurs. Inspections involve a critical, constructive examination of the workplace condition. Effective inspection can determine conditions that need to be corrected or improved to bring operations up to acceptable standards, both from a safety and an operational standpoint. An important side benefit of inspections is an improvement in operations that ultimately results in increased efficiency, effectiveness, and profitability.

Inspections must be viewed positively, as they are an important part of a facility's overall hazard identification and control program. If you eliminate hazards, you eliminate accidents. Similar to the approach taken for accident investigation (see Chapter 4), emphasis must be placed on fact-finding, not faultfinding. Faultfinding leads to criticism. People being inspected in any department become defensive when an inspector or inspection team takes a critical approach to the walk-through.

Policy and Planning for Inspection Programs

Inspection programs take careful planning. Policy must be established that makes inspections and audits part of the overall facility safety management system. A detailed plot plan is needed to identify location of facilities and equipment that make up the functioning whole of the site. Codes and standards must be gathered/written that define the criteria for what is safe and/or what is not safe. A family of inspection types, frequencies, participants, and procedures need to be in place. Measurement and monitoring systems should be defined for chemical exposures that may have toxic effects on employees. Inspectors need forms and methods to record and report hazards. Follow-up procedures are needed for corrective action to remove or reduce hazards discovered during inspections.

Hazardous Conditions to Inspect

Hazardous conditions in the workplace are everywhere. Which of the following examples may be found in your plant?

- Poor housekeeping in a paint room or a chemical process area
- Inadequate general ventilation and local exhaust for spray booths
- Wrong fire extinguishers for type of fire expected and/or improperly maintained
- Improper or no hazardous material container labeling
- Improper or lack of markings for pipes containing solvents or chemicals
- Blown or improperly sized rupture disks that have not been replaced
- Electrical fixtures in hazardous areas that are not properly rated
- Other ignition sources near the presence of flammable vapors
- Lack of in-rack sprinklers in a flammable/combustible liquids warehouse
- Lack of emergency information postings, including evacuation routes
- Leaks of toxic materials in pipe fittings and pumps
- Potential overpressure due to runaway chemical reaction

These hazards are accidents waiting to happen. Determine the categories of what to inspect or study by using the generic categories found in Figures 5-1, 5-2, and 5-8 as a guide.

Inspection Types, Frequencies and Participants

Under OSHA's general duty clause, management has the responsibility for providing a safe workplace. Thus, management needs to be involved in inspections and participate in the responsibility for finding hazards and carrying out the work of corrective action.

Safety and health professionals recommend facilities establish a "family" or system of inspections and audits (these terms are often interchangeable). Your system will be most effective when it involves the mix of inspection types that best meet your needs.

Most safety coordinators will relate to the following work condition inspection categories

- Formal audits
- Regulatory inspections
- Loss prevention surveys
- Informal inspections
- Formal self inspections

Formal Audits and Regulatory Inspections

External professionals or corporate staff may conduct formal audits. Some participating personnel may be specially trained, certified, and/or licensed.

Examples of regulatory inspections are those conducted by OSHA and EPA. OSHA inspections will almost always follow a fatality or hospitalization of 3 or more employees. Employee complaints may also trigger an OSHA visit if the hazard represents imminent danger. OSHA provides much information relative to agency inspections and audits on their web page.[2] Helpful topics include inspection priorities (imminent danger, catastrophes and fatal accidents, complaints, programmed and follow-up inspections). Information is also available regarding inspection of facility programs (including air contaminants and exposure controls, blood borne pathogens, confined space entry, chemical hygiene plan, emergency planning and response program, flammable liquids storage and handling, hazardous waste programs, personal protective equipment, respiratory protection, hazard communications, spray-finishing operations and others). Following an inspection, OSHA may issue citations without penalties, issue citations with proposed penalties that must be approved by state or regional officials, or determine that neither is warranted. Penalties range from other-than-serious (up to $7000 for each citation) to willful (minimum penalty of $5000 each and up to $70,000) violations. Note that OSHA inspectors prefer to cite standards right from the current code, for example 29 CFR1910.1200 Hazard Communication. However, if there is an unsafe condition and no standard exists, the general duty clause may be used as a basis for citation (Figure 8-1). There are four requirements for use of the general duty clause:[3]

1. A hazard existed and,
2. The hazard was recognized and,
3. The hazard could lead to serious injury or death and,
4. Feasible ways exist to abate the hazard.

Requirements 2 and 4 reinforce attention to immediate corrective action following self-inspections.

Organizations often conduct internal formal audits that may or may not include a physical inspection. For example, OSHA requires process safety audits every 3 years. These audits are best staffed by representatives from a number of plants, including persons trained in PSM audit protocol. The scope may include a single or all EHS (environmental, health and safety) programs and the audit team is best led by a divisional or staff EHS manager. Findings are usually made using interviews and document review and then measured against requirements of a formal standard.

Loss Prevention Surveys

Corporate safety specialists frequently conduct annual or semi-annual inspections under the heading of Loss Prevention Survey (LPS) or a similar name. This type of inspection normally has a comprehensive scope and takes 2–4 days or more. The inspector will usually audit the OSHA log in detail, examine written safety and health policies and procedures, interview employees, and conduct a walkthrough inspection of facilities and equipment.

Findings should be measured against specific regulatory standards if at all possible. A preliminary report is made before the inspector leaves the plant. A formal report that usually follows within two weeks is distributed to division and corporate management as well as the legal department.

Sometimes an LPS will result in more detailed inspections and tests. The follow-up may involve mechanics, electricians, and maintenance personnel that are trained, certified, or licensed technicians. Detailed inspections are often mandated by federal/state/local codes or standards, and may involve specialized equipment such as boilers and other pressure vessels, process control systems, cranes, hoists, and fire extinguishers. An example of detailed work is specific equipment inspections and tests like those necessary to meet mechanical integrity requirements for OSHA's Process Safety Management regulation. It often makes sense for facilities to initially hire outside experts for this task and then undergo the work to certify or license their own people.

When a potential health hazard is involved, industrial hygienists may also participate or follow-up on an inspection and do detailed work such as:

- Sample air for toxic vapors, gases, radiation, and particulates
- Test materials for toxic properties
- Measure and test ventilation and exhaust systems
- Measure physical stresses like noise, heat, and radiation

Follow-up on corrective action is a plant responsibility. A common practice is for the plant to communicate quarterly or semi-annual status reports to a function like corporate safety and/or the legal department.

Informal Inspections and Reviews

Informal inspections are most often conducted by internal personnel. An example is a routine department housekeeping inspection, or before a VIP visit. This kind of simple "walkthrough," may be general (plant-wide and all hazard categories) or specific (concentrating on one or two items like flammable liquids and ignition sources).

Normally, informal inspections involve one person, or at most, a few people. The report may be oral or preferably written.

A Responsible Care® review (see Chapter 11) is a good example of an informal review. Here again, interviews are held and documents are examined. In the case of Responsible Care, progress is measured according to categories like evaluating, implementing, practice-in-place, etc.

Self-Inspections and Employee Participation

Facility self-inspections are clearly the most important member of the inspection family. They may consist of informal inspections (recommended weekly for departments), followed by formal, monthly, plant-wide inspections. It is a good idea to conduct the plant inspection just before the safety committee meeting to encourage prompt corrective action. Companies experienced in continuous safety improvement frequently appoint employees to direct a portion of their time to safety and health for periods of a year or two. This service includes coordinating self-inspections.

Consider including the following functions:

- Lead operators and/other hourly employees from the same and other departments than the one being inspected
- Process, facility, and maintenance engineering
- Safety, health, and environmental coordinators
- Quality assurance chemists, engineers, and technicians

Department level inspections frequently do not follow a set routine and usually no formal checklist is used. Critics complain that informal self-surveys are haphazard and miss too much. However, they often involve the right mix of people, so the sense of ownership and involvement is high.

Be sure to involve people from a number of departments during plant-wide inspections. Those that work in a given area every day are prone to overlook many unsafe work conditions. It is good to invite co-workers or even outsiders into self-inspections in order to gain a new perspective.

Inspectors need to focus on physical, chemical, biological, and other hazards (Figures 5-1 and 5-2) regardless of whether a checklist is used. Corrective actions should be initiated immediately and status reviewed during the monthly safety and health committee meeting.

Hazards and Exposures

Workplace inspections should concentrate on actual or potential hazards that cause worker exposure. For example, suppose the inspection team comes across a parts-degreasing tank filled with acetone. A team member responds to an employee complaint that the odor is overpowering and that he is afraid of fire caused by passing fork lifts. The inspector notices that the hood over the tank does not draw much of the vapor away and holds a handkerchief up to the vent to prove it. The observation is noted on the clipboard inspection form and the leader explains that someone will check the complaint against OSHA requirements and report their findings to the employee. In this case, 1910.124 defines the general requirements for dipping and coating operations. Ventilation must keep the airborne concentration of the solvent below 25% of the lower flammable limit (LFL). Ventilation must be provided to maintain the 8-hour, airborne, time-weighted average concentration of acetone less than 1000 ppm. Air monitoring in the employee's breathing zone (Figure 1-2) for an entire shift will determine whether or not the exposure limit has been exceeded.

Inspection Procedures

It is useful for the facility safety manager to establish a process or procedure for self-inspections. Although the actual description and sequence of steps depends on the situation, safety managers should generally set up a simple procedure such as the following:

1. Prepare to inspect
2. Conduct the walk-through and record hazards
3. Write the inspection report
4. Follow-up system for corrective action

Prepare to Inspect

Gather tools for the inspection including clipboards, inspection forms, pencils, tape measure, ruler, calipers, micrometers, gauges, flashlight, etc. Depending on the type and depth of inspection, consider the following equipment for the inspection kit:

- Camera and tape recorder
- Electrical testing equipment
- Air sampling equipment and devices to measure temperature, noise, and light
- Sample containers
- Personal protective equipment required in the area

Using the site layout, divide the entire facility into sections that correspond to areas of responsibility. Include operating department locations, yards, buildings, equipment, forklifts, vehicles, hoists, and other items from Figures 5-1 and 5-2.

It is important to schedule the inspection at a time when the plant is operating normally. Establish the inspection route in advance. Plant-wide inspections should generally follow the manufacturing process from beginning to end. Be sure to include loading/unloading docks and tank farms. In other situations (for example, in batch chemical plants), the inspection will proceed from building to building.

Team members should meet to decide who is going where and if some members will split off to concentrate on specific areas. The leader may assign specific hazard categories or inspection items to individuals having expertise in those areas.

Prior to the walkthrough, review records of accidents that have occurred in the areas to be inspected, and see if corrective action has been taken and if it is working effectively.

Use of Material Safety Data Sheets

Using sections of the appropriate MSDS sheet for each hazardous chemical used, a team can develop a list of items for department or plant-wide inspections.

Example—Anhydrous Ammonia

Consider the MSDS for ammonia (Figure 1-1), and hazardous conditions to look for during the walkthrough:

1. Chemical Product and Company Identification
2. Composition/Information on Ingredients
3. Hazards Identification
   - Are there places in the plant where the liquid/gas is under pressure?
   - Where are the most likely places for loss of containment?
   - Where have we had previous incidents involving personnel exposure to ammonia?
   - Do we have ammonia monitors and if so, where?
4. First Aid Measures
   - Do we have an ammonia kit in the first aid packet?
5. Firefighting Measures
   - Are there ammonia storage areas in the plant exposed to external fire?
   - Are there provisions for water spray or other means of cooling tanks or cylinders?
   - Are there nearby ignition sources that would be a problem with loss of containment?
6. Accidental Release Measures
   - Are there confined spaces in the plant that ammonia can enter?
   - Are there provisions for water deluge or other means of reducing vapors?
   - Are there provisions for isolating and disposing ammonia-containing hazardous waste?

7. Handling and Storage
   - Are ammonia cylinders secured upright with screw valve caps in place?
   - Are full and empty cylinders stored separately?
   - Is cylinder storage in a place where the temperature will not exceed 125 deg. F?
8. Exposure Controls/Personal Protection
   - Are there adequate provisions for general ventilation in these areas?
   - Is local exhaust provided to maintain ammonia concentration below the TLV?
   - Is respiratory protection provided where needed that conforms to OSHA 1910.134?
   - Are gloves and eye/face protection provided as required?
9. Physical and Chemical Properties
10. Stability and Reactivity
    - Do the ammonia handling procedures avoid oxidizing agents, halogenated compounds, acids, and other incompatible materials?
11. Toxicological Information
12. Ecological Information
13. Disposal Considerations
    - Are ammonia wastes kept from contaminating surroundings on and offsite?
    - Are unused quantities returned to the supplier in the original container?
14. Transport Information
    - Are cylinders transported in a secure position and in a well-ventilated area?
    - Are containers properly marked per DOT standards?
15. Regulatory Information
    - Are all applicable federal, state, and local regulations being followed by the facility? By suppliers? By customers?

The reader can develop his or her own list of inspection items using the MSDS for specific chemicals handled in the plant.

Making Observations and Recording Hazards

Inspectors should contact the department head or area supervisor for a briefing before beginning the walkthrough. The person in charge may want to participate in the inspection (this helps establish department self-inspections). If they do not participate (or if their participation is against company policy), be sure to brief them regarding the team's findings. This includes a discussion of each recommendation for particular hazards or unsafe conditions.

An objective inspector relationship with the area supervisor, employees, and other team members is critical to objectivity. Do not override the responsibility of the supervisor. For example, if an inspector sees an employee deviating from established safe work practices, ask the supervisor, not the employee, for clarification.

Inspectors can ask questions about operations unless company policy or department rules prohibit conversation with employees. Whatever approach is decided upon, take care not to override authority of the supervisor. It is his or her job to assure safe work procedures. It is the inspector's job to inspect and report.

Watch, ask about, or imagine workers interacting with facilities and equipment. Ask yourself: "What's wrong with this picture?" In fact, a picture is "worth a thousand words" when it comes to describing an unsafe work condition.

Make detailed notes regarding obvious or suspected problems. Identify the building, room, and equipment (Figure 5-3). The inspection team should ask:

- Where is the problem?
- What is the hazard?
- What is the applicable requirement for a safe work condition? (This may be determined later by looking up applicable standards.)
- What is the appropriate method of hazard control?

For tracking corrective action, set up a numbering and code system to distinguish between temporary and permanent corrections. For example, you may choose to circle the items in Figure 5-3 where intermediate safety measures need to be taken. Once a permanent fix is in place, the item is crossed off, or an "X" is placed next to the item number. This practice greatly facilitates preparation of the inspection report.

Inspection Forms and Reports

Without a written report, there is little value to the inspection. Inspection reports from each department will be reviewed during monthly safety committee meetings as well as serve as a basis for corrective action. Keep the inspection findings on the list and update the status until the item can be marked complete.

There are many types of inspection reports and the best format is one that facilitates corrective action. A checklist (Figures 5-1 and 5-2) may be used by department personnel to check-off compliance and mark deviations. Most checklists are too comprehensive for any one person to follow during an inspection. Inspection teams typically divide up the categories or ask each participant to generate their own list of inspection items.

Physical Safety
Check for the presence of the following:

Electrical Hazards	Non-ionizing Radiation
Fire Extinguishers	Ionizing Radiation
Confined Spaces	Infrared Devices
Fall Hazards/Egress	Lasers
Heat	Intense Visible Light
Noise	Magnetic Fields
Vibration	Radio Frequency Waves
Repetitive Motion Hazards	Microwaves
Lifting Hazards	X-Ray Producing Machines

Chemical Safety
Check for the presence/storage of the following chemicals:

Acids	Cleaners/Solvents	Halogens
Adhesives	Compressed Gases	Hazardous Waste
Alcohols	Cryogenics	Inks/Dyes
Aldehydes	Cyanide compounds	Isocyanates
Alkalis	Epoxy Resins	Metal Fumes
Amines	Esters	Metals (heavy)
Ammonia	Explosives	Lubricants
Aromatic compounds	Flammables	Oxidizers
Asbestos	Freons (CFCs)	Paints
Chromates	Fuels	Pesticides

Biological Safety
Check for potential presence of the following:

Bloodborne Pathogens	Allergens
Food borne Pathogens	Fungus
Other Pathogens	Insects

Figure 5-1 Emery and Savely's Inspection Checklist[4]

Figure 5-2 is a checklist with a place at the end to record when and who produced the corrective action. In this figure, the electrical, confined space entry, fire protection, housekeeping, hot work, hazardous materials, fork trucks, and employee protection categories have all been modified to better represent facilities that handle chemicals and flammable/combustible liquids.

Company_____
Location: _____ Inspected by: _____ Date: _____

General

Ok

❑   Do you have an active accident prevention program?

❑   Do you have an active Hazard Communication program?

❑   Do you have an active Process Safety Management program?

❑   Do you have an active contractor safety program?

❑   Do you have written procedures for confined space entry, hot work, lock-out-tagout, and line breaking?

❑   Other written programs?

Workplace

Electrical Wiring, Fixtures and Controls

❑   Are workplace electricians familiar with the requirements of the National Electrical Code (NEC, NFPA 70)? Are all installations in accordance with listing or labeling? (110-3)

❑   Do you specify compliance with the NEC (especially article 500- Hazardous Locations) for all contract electrical work?

❑   If you have electrical installations in hazardous dust or vapor areas, do they meet the NEC Article 500 requirements for hazardous locations?

❑   Does all wiring in hazardous areas conform to class I, II, III/division 1 or class I, II, III/division 2 requirements according to NEC article 500?

❑   Does someone competent in the NEC check your electrical system periodically?

Walkways, Stairways, and Exits

❑   Are all aisles and passageways kept clear?

❑   Are standard stair rails or handrails provided for all stairways having four or more risers? Is the handrail capable of withstanding a downward load of 200 pounds?

Figure 5-2 Self-Inspection Checklist[5]

❑ Are steps provided with a surface that renders them slip resistant? The slip resistant material in good condition?

❑ Are all exits visible and unobstructed?

❑ Are all exits marked with a readily visible sign that is properly illuminated? Are exit paths marked, illuminated, and free from obstruction? Are exit aisles sufficient in number and width?

❑ Are exit doors equipped with crash bars or other means to facilitate quick escape?

❑ Are there sufficient exits to ensure prompt escape in case of an emergency?

❑ Are exits and exit paths free from storage of flammable or combustible materials?

❑ Do you take special precautions to protect employees during construction and repair operations?

❑ Are outside walkways visibly marked, especially around vehicle roadways?

Confined Space Entry

❑ Are areas with limited occupancy posted as a *confined space* and is access/egress controlled to persons specifically authorized to be in those areas?

❑ Are confined space entry responsibilities in place for the entry supervisor, attendant at-point-of-entry, person making the entry, and the emergency rescue team?

❑ Does your confined space entry system include training, use of a permit, lockout/tagout procedures, lifelines, and equipment for hoisting, temperature, oxygen and vapor or gas check, alarms and communication, respiratory protection, and standby rescue?

Fire Protection

❑ Are portable fire extinguishers provided in adequate number and type?

❑ Are fire extinguishers inspected monthly/quarterly for general condition and operability and noted on the inspection tag?

❑ Are fire extinguishers recharged regularly and properly noted on the inspection tag?

❑ Are fire extinguishers mounted in readily accessible locations? Are they provided near flammable/combustible liquid dispensing pumps?

❑ If you have interior standpipes and valves, are these inspected regularly?

❑ Is the fire alarm system tested at least annually?

❑ Are plant employees periodically instructed in the use of extinguishers and fire protection procedures?

Figure 5-2 Self-Inspection Checklist[5] (continued)

❏ If you have outside private fire hydrants, were they flushed within the last year and placed on a regular maintenance schedule?

❏ Are fire doors and shutters in good operating condition? Are fusible links in place?

❏ Are fire doors and shutters protected against obstruction? Are they marked "Do not block open?"

❏ Is there adequate clearance around flammable gas/liquid storage tanks and is there adequate protection from external impact? Is the LP/natural gas meter adequately protected?

❏ Is the fire lane free from obstruction?

❏ Is your local fire department well acquainted with your plant, location and specific hazards?

Automatic Sprinklers

❏ Are water control valves, air and water pressures checked weekly/monthly?

❏ Are control valves locked open? Is a tamper switch provided?

❏ Is maintenance of the system assigned to responsible persons or a sprinkler contractor?

❏ Are sprinkler heads protected by metal guards where exposed to mechanical damage?

❏ Is proper minimum clearance maintained around sprinkler heads?

Housekeeping and General Work Environment

❏ Is smoking permitted in designated "safe areas" only?

❏ Are NO SMOKING signs prominently posted in areas containing combustibles and flammables?

❏ Are covered metal waste cans used for oily and pain soaked waste?

❏ Are waste containers provided and emptied at least daily?

❏ Are paint spray boots, dip tanks, etc., and their exhaust ducts cleaned regularly?

❏ Are stand mats, platforms or similar protection provided to protect employees from wet floors in wet areas of the facility?

❏ Do your toilet facilities meet the requirements of applicable sanitary codes?

❏ Are uniforms provided for employees working with hazardous materials?

❏ Are washing and/or showering facilities provided?

Figure 5-2 Self-Inspection Checklist[5] (continued)

Welding, Flame Cutting, and Other Hot Work

❑ Are only authorized, trained personnel permitted to use such equipment?

❑ Is a permit required before hot work begins? Have hot work operators been g given a copy of operating instructions and asked to follow them? Do they have the proper protective clothing and personal protective equipment?

❑ Are welding gas cylinders stored so they are not subject to damage? Are valve protection caps in place on all cylinders not connected for use?

❑ Is a fire extinguisher provided at the hot work site?

❑ Do your procedures call for a fire watch where appropriate?

Hazardous Materials

❑ Are all materials used in your plant checked for toxic and hazardous properties as found on the Material Safety Data Sheet?

❑ Are there hazardous material warnings posted where exposure could occur?

❑ Do you have warning and signaling devices for toxic and flammable gases and vapors?

❑ Do you routinely do air sampling in areas where exposure is known to occur?

❑ Have appropriate control procedures such as ventilation systems, enclosed operations, safe handling practices, proper personal protective equipment (e.g. respirators, safety glasses with side shields, and/or goggles and face mask, and gloves, etc.) been instituted for toxic materials? Are control systems routinely maintained?

❑ Are all storage tanks and containers marked with appropriate hazard warnings?

❑ Are approved safety cans or other acceptable containers used for handling and dispensing flammable liquids?

❑ Are all flammable liquids that are kept inside buildings, stored in proper storage containers or cabinets? Are in-rack sprinklers provided for high rack storage of packaged flammables?

❑ Is there adequate sprinkler protection for dispensing flammable liquids?

❑ Do you meet OSHA standards for all spray painting or dip tank operations using combustible liquids?

❑ Are oxidizing chemicals stored in areas separate from all organic material except shipping bags?

❑ Is ventilation equipment provided for removal of air contaminants from operations such as production grinding, buffing, spray painting and/or vapor degreasing, and is the system operating properly?

Figure 5-2 Self-Inspection Checklist[5] (continued)

❑ Are diking and/or absorbent materials available to prevent hazardous materials spills from entering sewers and waterways?

❑ Are protective measures in effect for operations involved with X-rays or other radiation?

Fork Lift Truck Operations

❑ Are all diesel, electric, gasoline, LP Gas, or other fueled trucks of the type required by OSHA 1910.178 for Class I, II, and III hazardous areas?

❑ Are only trained personnel allowed to operate forklifts?

❑ Do you require completion of a daily/weekly inspection checklist?

❑ Is overhead protection provided on high lift rider trucks?

Employee Protection

❑ Is there a hospital, clinic or infirmary for medical care near your business?

❑ If medical and first aid facilities are not nearby do you have one or more employees trained in first aid?

❑ Are your first-aid supplies adequate for the type of potential injuries in your workplace?

❑ Are there quick water flush facilities available where employees are exposed to hazardous materials?

❑ Are hard hats provided and worn where any danger of falling objects exists?

❑ Are protective goggles or glasses provided and worn where there is any danger of flying particles or splashing of corrosive materials?

❑ Are approved respirators provided for regular or emergency use where needed?

❑ Is all protective equipment maintained in a sanitary condition and readily available for use?

❑ Where special PPE is needed for workers handling chemicals and is it available?

❑ When lunches are eaten on the premises, are they eaten in areas where there is no exposure to toxic materials, and not in toilet facility areas?

❑ Is protection against the effects of occupational noise exposure provided when the sound levels exceed those shown in Table G-16 of the OSHA noise standard (90 DbA for 8 hours)? Are employees routinely tested for hearing loss?

Repairs/corrections must be completed by (date) _____
Report routed to _____ (date) _____
Repairs/corrections have been completed by _____

Figure 5-2 Self-Inspection Checklist[5] (continued)

The danger of a checklist is the tendency to simply check off items without a reasonable examination of actual and potential hazards. A word of caution about this "check-the-box" approach to inspection: plant management often complains about checklists because employee inspectors focus on getting through the list without finding hazards. Asking "what's wrong with this picture?" and the other key questions raised above is critical to an effective outcome. Most experienced inspectors use a blank tablet to record observations and then simply type up a list and location of unsafe work conditions.

INSPECTION REPORT

Area Inspected Plant A   Date and Time of Inspection 11/30/01 11:00 a.m.
Inspector and Title Ron Van Safety, Hazard Control Specialist   Date of Report 12/15/01
Report Submitted to: Bob Blazer, General Manager, Loren Chemise, Operations, file

*No. of items carried over From previous report 3		No. of items added to this report 4				Total No. of items on this report 7	
Item	**Hazard Classification**		**Hazard Description**	**Specific Location**	**Supervisor**	**Corrective Action Recommended**	**Corrective Action Taken**
	Consequences	Probability					
*1	II	C	Wrong fire extinguishers	Distillation	J. Jones	Purchase (6) FFFPs model 252-ethanol specific	
*2	III	B	hot slurry tank lid	Mash operations	B. Smith	Purchase "Hot Surface –Wear PPE" label	
3	II	C	Extension cord on ground	Near large ammonia tank	A. Anderson	Replace with permanent conduit	
4	II	A	Ammonia leak- **immediate repair**	Ext. ammonia tank piping	P. Johnson	Repair flange, relocate remote shutoff valve closer to tank	
5	II	A	Caustic spill between tank and dike	Fermentation operations	O. Olsen	Contract hazmat cleanup responder	
*6	III	A	Hot steam trap, condensate pump at floor level	Process area 2	S. Peterson	Insulate from personnel exposure	
7	III	C	No dirt covers on eye wash stations	Plant	D. Nelson	Purchase, clean stations and install	

Consequences
- I. **Catastrophic-** may cause death or loss of a facility.
- II. **Critical-** may cause severe injury, severe occupational illness, or major property damage.
- III. **Marginal-** may cause minor injury or minor occupational illness resulting in lost workday(s), or minor property damage.
- IV. **Negligible-** probably would not affect personnel safety or health and thus, less than a lost workday, but nevertheless is in violation of specific criteria.

Probability
- A. Likely to occur immediately or within a short period of time when exposed to the hazard.
- B. Probably will occur in time
- C. Possible to occur in time
- D. Unlikely to occur.

Figure 5-3 Inspection Report Format[6]

Figure 5-3 is more appropriate for identifying a hazard classification and what corrective action is recommended or taken.

The best way this format is to:

1. Describe the hazard, location, and supervisor.
2. Classify the hazard using the consequence and probability guidelines. Items marked IA or IIA are emergency classifications and should prompt immediate action.
3. Research the requirements for a safe work condition, if necessary, and recommend corrective action.
4. Use the last column to document the corrective action.

It may be necessary to research criteria for safe work conditions before writing the final report. For example, the inspection team may be unsure or the area supervisor may disagree over whether maintenance cleaners and degreasers need to be listed as hazardous materials. Here, check the CFR for guidance (29 CFR 1910.1200(e)(1)(i)).

When satisfied that the content is accurate, rate hazard priorities and complete the final inspection report.

Follow-up Corrective Action

A documented, yet uncorrected, unsafe work condition is a smoking gun! Company and regulatory audits that come on-site always look for a paper trail to see if deficiencies have been corrected. However, follow-up on inspections is an opportunity to put hazard identification and control and company policy before regulatory compliance. Remember that the benefits of safe work conditions include high morale and increased efficiency and effectiveness as well as accident prevention.

The inspection team or someone else may make assignments for correction action. However this is done, be sure to involve the area supervisor. As stated earlier, review corrective action status during monthly safety meetings. Close and document all completed items, and update the status of those yet to be completed.

Refer to the procedure for incident investigation (Chapter 4) as a good guide for taking corrective actions. The fix must address the root cause of an unsafe work condition, not just patch it up in a superficial fashion. Evaluate the change against the safety precedence sequence (Figure 4-6).

The shop work order or other means of authorizing repair needs to explain exactly how the solution is to be implemented. Determine whether faulty equipment can be operated until repairs are complete. The person responsible for completing the repair needs to adhere to start and finish dates (usually expressed in 30, 60, or 90-day intervals).

Chemical Exposure Sampling and Monitoring

OSHA is continuing to emphasize more accurate and complete workplace assessment. A good example is the recently revised standard for PPE which requires a hazard assessment prior to specifying PPE (such as respiratory protection).

Physical inspections of work stations involving chemical handling often trigger sampling and monitoring to quantify the exposure in terms of a permissible exposure limit (PEL). This potential hazard is found in many plant functions that include generally include operations, maintenance and laboratory. Review the discussion on air sampling and monitoring in Chapter 1.

Consider purchase or rental of direct reading instruments wherever practical, since they are easier to use. State health departments are also an inexpensive source for special instruments. Note that placement of sampling and monitoring devices is a job for a professional industrial toxicologist or hygienist. At the very least, plant personnel should be trained to place and use these devices. Larger companies often have a service group to analyze the samples and report the findings. Take care to assure that the sampling method is compatible with the method used for analysis.

Safe Work Conditions and Process Hazards Analysis

Many of the catastrophic events that have happened in the chemical process industry would have been prevented if more attention had been paid to safe work conditions. The Pasadena, Texas and the Bhopal, India accidents may have been prevented by application of process safety management (PSM) and risk management principles (RMP).

Both events could have been predicted using process hazards analysis (PHA) as a tool for identifying hazardous work conditions.

Procedures for Completing PHA Studies

This section outlines a procedure utilizing PHA and involves other elements called the "engine of process safety management." These elements are:

- Process Safety Information
- Process Hazards Analysis
- Operating Procedures
- Pre-start Safety Review
- Management of Change

OSHA's Process Safety Management Standard (29 CFR 1910.119) and the EPA Risk Management Planning regulation (Clean Air Act Section 112 (r)/40 CFR Part 68) require much of facilities that store or handle chemicals. The purpose of each regulation is to prevent or minimize the consequences of catastrophic releases of toxic, reactive, flammable, or explosive chemicals that may result in toxic, fire, or explosion hazards. Each agency defines process broadly and has their own list of chemicals and threshold quantities for coverage. However, there is much overlap.

The procedure for completing PHAs involves 10 steps:

1. Choose a process safety coordinator.
2. Prioritize the PHA list.
3. Define the PHA scope and purpose.
4. Form the PHA team.
5. Schedule and conduct the PHA.
   a. Define process nodes or sections.
   b. Analyze process or procedure deviation.
   c. Choose consequences of interest.
   d. Determine potential cause of deviation.
   e. Examine safeguards and consequences of failure.
   f. Determine action items.
6. Follow-up and resolve all action items.
7. Update operating and safety information.
8. Finalize Management of Change policy and conduct Pre-start Safety Review.
9. Issue the PHA report and communicate to affected employees.
10. Revalidate each PHA every five years.

Step 1—Choose a Process Safety Coordinator

Choose a process safety coordinator during initial implementation of process safety management. Do not expect the facility safety coordinator to do the process work in addition to his or her regular duties. Pick a key process engineer, technical manager, or someone that is leading your quality improvement effort. This job may be best when assigned for one or two years, then rotated to someone else.

Step 2—Prioritize the PHA List

The OSHA regulation lists cover processes containing 139 toxics and flammables (excluding those used for comfort heating) in quantities greater than 10,000 lbs. In fact, there are two such lists. Because PSM is the prevention plan for EPA's Risk Management Plan, we must also consider the EPA chemical list.

The current EPA list contains 77 toxics and 63 flammables. It is difficult to rationalize why some chemicals appear on OSHA's list but not on the EPA list and vice versa. When deciding the order for the PHAs, you must demonstrate a priority rational that takes into account the extent of process hazards, the number of potentially affected employees, and the age and operating history of the process. There is no formula for this process. A facility will usually have no trouble rationalizing the order for their list.

Step 3—Define the PHA Scope and Purpose

It is important to determine the scope of the study by deciding which specific process areas to include. Choose appropriate process flow diagrams or piping and instrumentation drawings (P&IDs) and the associated operating procedures. Decide what unwanted consequences will be considered. The study should cover, at a minimum, process hazards that will potentially result in large-scale fires, explosions, and toxic releases. Many teams will also address personnel-type hazards resulting in chemical exposure, slip/trip/fall, struck-by machine or object, material handling, etc. (However, this is not the main purpose of the study.)

Step 4—Form a PHA Team for Each Covered Process

Because process safety involves the process and equipment that makes the product, manufacturing must be involved in safety as well. Select a team of people from the following functions:

- Process safety coordinator
- Process engineering
- Division and/or resident (equipment) engineering
- Plant or maintenance engineering
- Operations (chemical operators and manufacturing) supervisors
- Personnel safety coordinator
- Process instrumentation and control systems engineering (PI&CS)

The PSM coordinator and PHA teams will also be heavily involved in other process safety activities, including involving employees, writing operating procedures, gathering and updating process safety information, conducting process safety training, writing mechanical integrity procedures, and developing management of change procedures.

The team leader should be trained in the use of PHA methodology and not be involved with the process itself. This person may come from the plant, corporate office, or outside the organization.

Step 5—Schedule and Conduct the PHA

The leader will define the process sections (sometimes called nodes) to be included in the study. Using the HAZOP method, for example, he or she will draw circles around the vessels in Figure 5-4.

The solvent and raw material storage tanks, reactor, reflux condenser, solvent stripper, etc. will each be a separate node.

The team will then consider abnormal deviations for each line coming in and out of the vessel as well as deviations within the vessel. In the next example (Figure 5-5), deviations within a chlorine vaporizer are examined.

In Figure 5-5, high and low deviations in level, temperature, pressure, etc. are each evaluated in turn. Always consider the consequences of each deviation before brainstorming the possible causes. Note in item 7.1 (high level), over-chlorination in the bleach process can lead to a small chlorine release. However, high temperature (item 7.3) is the real concern, since an exothermic reaction between chlorine and the vaporizer materials could release a significant amount of chlorine to the atmosphere. Therefore, the team should study the possible causes of high temperature in detail.

The next step is to consider existing safeguards (temperature indicators and a high temperature alarm in the control room) and the possibility of failure. At this point, the team discusses the adequacy of the safeguards.

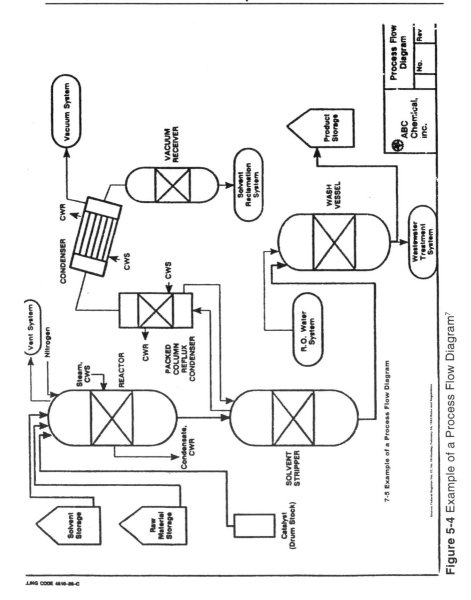

Figure 5-4 Example of a Process Flow Diagram[7]

For the chlorine vaporizer example, the team recommended five action items related to high temperature and three actions for low temperature. Note that the wording is to "consider" the action, rather than "replace" or "add" safeguards. This approach is important in order for team members to have an opportunity to study the action item before the team makes a recommendation to management.

Company: JBF Paper Company	Plant: Pulp Mill	Site: Anywhere USA	Unit: Bleach Process	System: Railcar Unloading System
Method: HAZOP	Type: Heat Exchanger	Design Intent: Vaporize chlorine		

Number: Drawing: D.20.03.F.100.17, D.20.03.P.001, **Procedure:** SOP-03-001, SOP-03-003

Team Members: Woody Davis, Kathryn Grady, Alex James, Marcus Samuels, Richard Thomas

No.: 7	Description: CHLORINE VAPORIZER

Drawing: D.20.03.F.100.17, D.20.03.P.001, **Procedure:** SOP-03-001, SOP-03-003

Item	Deviation	Causes	Consequences	Safeguards	Action Items
7.1	High level	Low temperature (linked from 7.4) High flow - Chlorine gas supply line for bleaching line A (linked from 8.1)	Potential carryover of liquid chlorine into the chlorine gas line and over chlorination in the bleach process (small chlorine release)		
7.2	Low level	Low/no flow - Chlorine unloading line (linked from 4.2)	Low pressure - Chlorine gas supply line for bleaching line A (linked to 8.9) High temperature (linked to 7.3)		
7.3	High temperature	Excessive superheat in the steam supply Operator setting steam pressure control valves incorrectly Steam pressure control valves failing to close or transferring open Low level (linked from 7.2)	Accelerated corrosion Exothermic reaction between chlorine and the vaporizer materials of construction, damaging the vaporizer and potentially releasing a medium amount of chlorine to the atmosphere High pressure (linked to 7.5) Tube leak or rupture (linked to 7.9)	Temperature indication and high temperature alarm in control room for the vaporizer that is indicated by the selector switch	28. Consider installing a de-superheater in the steam system for the chlorine vaporizers 29. Consider installing a pressure relief valve on the vaporizer steam jacket to help limit the steam pressure in the vaporizer shell (even if the control valves fail) 30. Consider using a more direct control scheme for vaporizing the chlorine liquid (e.g., controlling the temperature at the discharge of the vaporizer by regulating steam flow) 31. Consider replacing the manual selector switch in the high temperature alarm circuit with an automatic switch or with dedicated temperature alarm circuits for each vaporizer 32. Consider providing an interlock that would close the steam valve to the vaporizers on high temperature

Figure 5-5 Example PHA Worksheet[8]

Item	Deviation	Causes	Consequences	Safeguards	Action Items
7.4	Low temperature	Operator failing to open or inadvertently closing a valve in the steam system Operator setting steam pressure control valves incorrectly Steam pressure control valve failing to open or transferring closed Vaporizer tube fouling	Low temperature - Chlorine gas supply line for bleaching line A (linked to 8.7) High level (linked to 7.1)	Temperature indication and low temperature alarm in control room for the vaporizer that is indicated by the selector switch	30. Consider using a more direct control scheme for vaporizing the chlorine liquid (e.g., controlling the temperature at the discharge of the vaporizer by regulating steam flow) 31. Consider replacing the manual selector switch in the high temperature alarm circuit with an automatic switch or with dedicated temperature alarm circuits for each vaporizer 41. Verify that the excess flow valves in the railcar liquid dip tubes are designed to operate properly (i.e., restrict chlorine releases) even if both dip tubes are in use
7.5	High pressure	Low/no flow - Chlorine gas supply line for bleaching line A (linked from 8.2) High temperature (linked from 7.3)	High pressure - Chlorine railcar (linked to 1.5) Potential damage to the vaporizer if isolated from the relief valve on the chlorine railcar (linked to 7.9)	Local pressure indication	33. Consider providing a high pressure alarm for each vaporizer
7.6	Low pressure	Operator using the neutralization system eductor to remove chlorine from the vaporizer	Potential vaporizer collapse during chlorine evacuation through the eductor in the neutralization system, resulting in a small release of chlorine		5. Verify that all of the chlorine unloading and vaporizing equipment (particularly the chlorine railcars and vaporizers) can withstand the maximum vacuum created by the neutralization system. If the neutralization system can produce a vacuum capable of damaging equipment, consider modifying the neutralization system to minimize this potential
7.7	High concentration of water	Vaporizer not properly drained and purged after periodic maintenance High concentration of water - Chlorine unloading line (linked from 4.11) Vaporizer tube leaking while the vaporizer is out of service (linked from 7.9)	Tube damage resulting from acid corrosion (linked to 7.9) High concentration of water - Chlorine gas supply line for bleaching line A (linked to 8.10)	Vaporizer startup procedure	34. Consider requiring operators to check for water in the vaporizers before every startup

Figure 5-5 Example PHA Worksheet (continued)[8]

Item	Deviation	Causes	Consequences	Safeguards	Action Items
7.8	High concentration of nitrogen trichlorite	Distillation of nitrogen trichloride in the vaporizer	Potential explosion resulting in tube damage (linked to 7.9)		35. Consider implementing procedures to periodically remove nitrogen trichloride from the vaporizers
7.9	Tube leak or rupture	Corrosion/erosion Gasket, packing, or seal failure Improper maintenance Material defect High temperature (linked from 7.3) High pressure (linked from 7.5) Acid corrosion resulting from a high concentration of water (linked from 7.7) Detonation of nitrogen trichloride (linked from 7.8)	Medium release of chlorine to the atmosphere if the tube heads fail Small release of chlorine into the steam system and the sewer High concentration of water (linked to 7.7)	Capability to isolate manually Maintenance/operator response as required, including isolation if needed Video monitoring of the vaporizing area	36. Consider monitoring the steam condensate drain lines for chlorine
7.10	Shell leak or rupture		No consequences of interest		

Figure 5-5 Example PHA Worksheet[8] (continued)

Step 6—Follow-up and Resolve All Action Items

In this step, it is important to note OSHA's instruction to compliance officers regarding PHA findings and recommendations. Plant management is required to promptly address and resolve the team's findings, document the actions taken, and communicate these actions to affected employees. "Resolve" means that management either has adopted the recommendations or has justifiably declined to do so. When declining recommendations, management must document that one of the following reasons apply:

a) The team's basis for analysis that led to the recommendation contained factual errors.

b) The recommendation is not necessary to protect the health and safety of employees or contractors.

c) An alternate protective measure would provide a sufficient level of protection.

d) The recommendation is infeasible.

Step 6 represents an important handoff from the PHA team to plant management. The team should conduct their follow-up thoroughly and promptly. If recommendations are adequately supported, management generally can make a solid decision whether or not to provide the resources necessary.

Step 7—Update Operating and Safety Information

Once the team has completed the study, they may submit proposed changes to operating procedures and any other PHA-recommended action items. Since it must be maintained for the life of the process, PSI should also be updated to reflect changes resulting from the PHA.

Step 8—Finalize Management of Change (MOC) and Conduct Pre-Start Safety Review (PSR)

Pre-start safety reviews are required for new and modified facilities where changes have been made to PSI. Hold the meeting with both management and all members of the PHA team present. Date and document the results of the meeting, and place the records in the PHA or PSI file.

Note that the PSR must confirm that a management of change (MOC) system is in place and that modified facilities meet MOC requirements.

MOC is another element of process safety management that affects the safe work condition of equipment. OSHA requires written procedures for change (except replacements in kind) affecting:

1. covered process chemicals
2. process technology
3. facilities and equipment
4. work procedures

The documentation must include:

- Technical basis for the proposed change
- Impact on safety and health
- Modifications to operating procedures
- Necessary time period for change
- Authorization for the proposed change

Other management of change requirements involve training and systems for updating process safety information and operating procedures. The following section will concentrate primarily on changes to facilities and process equipment.

Changes to any process equipment or facility components need to go through appropriate safety and environmental reviews. The first step is to evaluate the change according to a simplified PHA format:

1. Deviation from the condition before the change
2. Possible adverse consequences
3. Potential cause(s) of the deviation
4. Existing and additional means of control

In order to save time and effort, consequences of interest are evaluated before causes are evaluated. Next, management gives the authorization to procure and install the new equipment. Training is conducted and process safety information is updated. Finally, a pre-start safety review is held, and the change is incorporated into the process.

Note that the system will not work if employees are not involved in establishing functional guidelines for change, procedures, and documentation. For example, the facility needs to install some form of the four-step procedure above. This process takes some time to complete. Meanwhile, the facility must establish some kind of MOC system that will authorize change safely and quickly for any time—day or night—that the covered process is operating. Full documentation can be completed the next day.

However, the MOC procedure must establish a technical person on-call or an alternative way to promptly review the impact of the change on safety and health. If satisfied, the authorizing person can either give the go-ahead or hold up the process until the system can be reviewed in more detail by the process safety team.

"Replacement in kind" means the replacement parts and equipment are the same as the original or current (at the time of the PHA) components. A centrifugal pump replacing a centrifugal pump is replacement in kind. If the replacement pump is a gear pump, the change must go through MOC. The process safety team is advised to define replacement in kind in terms of examples of specific process equipment components that do and do not meet the criteria.

Figure 5-6 is an abbreviated checklist of items that are sometimes overlooked by facilities. An example of a process change permit is found in Figure 5-7.

Facility Modifications

❏ Have buildings, trailers, etc., been constructed or relocated since the previous PHA such that they could be affected by a release?

❏ Have staffing levels changed since the previous PHA such that capabilities to quickly respond to emergency situations have diminished?

❏ Have traffic patterns (e.g. new rail spur, new chemical truck routing) changed since the previous PHA?

❏ Could the process be affected by a new release source?

❏ Can emergency responders reach the process easily, and are external impacts more likely than before?

❏ Are there new ignition sources near this process?

❏ Are new utilities being used to prepare equipment for maintenance?

❏ Has the area electrical classification changed such that some equipment is not properly rated for its service?

❏ Have facility modifications made alarms difficult to see or hear?

❏ Have facility modifications increased the number of locations that should have personal protective equipment (PPE) available?

Modifications to the Process

❏ Was a MOC system implemented before or in conjunction with the completion of the previous PHA?

❏ Have all PHA issues (e.g., facility siting, human factors) been addressed for equipment added to the process?

❏ Have equipment relief valves been changed from atmospheric discharge to a closed system (or vice versa)?

❏ Are new hazards present due to new operating modes?

❏ Have the hazards associated with equipment that has been returned to service since the previous PHA been evaluated?

Figure 5-6 Abbreviated Checklist for Management of Change[9]

Process	Department	Supervisor/Lead	Phone	Perm./Temp Change

Change Title	Number	Work order #	Diagrams/Drawings ❑ Yes Attached? ❑ No

Originator	Department	Date	Phone

Type of Change   __ major   __ planned
                 __ significant   __ emergency
                 __ minor   __ other _____

Description of Change:
_____
_____
_____
_____

Change to:
❑   Chemicals_____
❑   Process technology/procedures_____
❑   Equipment _____
❑   Facilities_____
Technical basis for change:_____
Impact on safety and health:_____
Modification to operating procedures? ___yes ____ no ___ SOP # ___ other #
Time period for change?_____

DOCUMENTATION UPDATES

Required	Complete?	Name-print	Initials	Date
Operating procedures		John Jones		
Safety procedures				
Master record				
Instrumentation				
Equipment database				
Reciprocal support agreements				
Training records				
Waste stream data base				
Storage tank inv.				
Air permitting				
Fugitive emissions				
APPROVALS_____		(list of approvals omitted)		

Figure 5-7 MOC Permit [10]

Note that the form in Figure 5-7 includes a description of the change, necessary approvals, and required documentation updates.

Step 9—Issue PHA Completion Report and Communicate Results to Affected Employees

A written plan for employee participation is the first PSM requirement listed by OSHA. Employees and their representatives must be consulted regarding development of PHAs and other PSM elements. In addition, they are to have access to PHA studies and the results. Normally, this is accomplished by choosing employees involved in the process to be a part of the PHA team and communicating results to affected employees.

Step 10—Revalidate Each PHA Every Five Years

The reader may remember that OSHA required 25% of all initial PHAs to be completed by May 26, 1994 and all PHAs to be completed by the same date in 1997. This means that the period May 1999 to May 2002 was a key period for PHA revalidation. Of course, there are many covered facilities that have just begun working on their PHA list and others that have not started at all. A recommended six-step process for revalidating PHAs is to:[11]

1. Review all modifications made to the process since the previous PHA

The purpose of this step is to verify that the revalidated PHA is current with respect to all process hazards as well as engineering, administrative, and other means of hazard control. If the MOC system has been maintained, revalidation simply means reviewing all completed MOC forms and folding them into the revised PHA. Unfortunately, many plants have insufficient documentation to support a hazard evaluation of each change made since the previous PHA; such plants must start with a new PHA rather than attempt a revalidation based on incomplete information and records. It is also a good idea to estimate the time required to review each change.

2. Review previous incidents

The PHA revalidation team must also review all relevant incident reports since the last PHA that had a *potential* for catastrophic consequences. This is another case where documentation may be lacking or non-existent.

The purpose of this review is to determine first if there is any trend to indicate that a hazard studied during the last PHA has a greater risk than was thought at the time. The second purpose of incident review is to catch hazards that may not have been considered hazards the first time. Regardless of purpose, reassessment at some level is necessary.

Allow about 10 minutes each to review incident reports without recommendations or those that have MOCs. For reports without MOCs, the time ranges from 5 minutes for small changes to over an hour for large and complex modifications.

3. Review the status and resolution of previous PHA recommendations.

This task is not specifically required by OSHA but has the benefit of assuring that all changes from step one have been identified.

4. Update the human factors or facility siting analysis.

Depending on whether or not these items were considered during the previous PHA, this step could require substantial time to complete. Address human factor topics including HAZMAT labeling, access to equipment and controls, display and control feedback, employee workload and stress, training, effectiveness of operating and maintenance procedures, and housekeeping issues. Siting considerations include unit layout and equipment spacing; location of large hazmat inventories; placement and construction of control rooms and motor control centers; location of possible ignition sources; location of highly occupied on and off-site buildings and nearby neighbors; firewater mains and backup water locations; location and effectiveness of drains, spill basins, dikes, and sewers; location of emergency stations such as showers, respirators, and other PPE; and electrical classification in and near hazardous locations.

5. Address hazards associated with non-routine operating modes.

A disproportionate number of accidents occur during non-routine operations such as start-up, shutdown, maintenance, and sampling. These modes need to be evaluated during the PHA revalidation. A modified HAZOP or "what-if" analysis is a good approach to identifying these types of hazards.

6. Ensure that the PHA meets the requirements of any existing or new regulations, industry standards, or internal company requirements.

Finally, the team needs to assign someone to review current regulations for new requirements adopted since the last PHA. A good example is that EPA's RMP program requirement calls for worst-case and alternative offsite consequence analysis (OCA). The last PHA may have conformed only to OSHA requirements. Catastrophic events such as toxic release, fires, and explosions may have been a part of the initial study, but in this case offsite impact may not have been quantified.

For complete guidance on PHA revalidation, consider purchasing the CCPS book, *Revalidating Process Hazards Analyses*.[12]

Hazards and PHA Methodology

The intent of the PHA is to identify and control process hazards. A good starting point is to look for unsafe forms and levels of energy. For example, materials that form toxic vapors, are flammable, thermally unstable, shock sensitive, and/or capable of self-polymerizing are common hazards that typically cause unwanted events. Also look to elevated temperature, pressure, level, and volume relationships that can release energy and lead to loss of chemical containment.

OSHA provides a good deal of leeway in PHA methodology. OSHA requires the study to be appropriate given the complexity of the process and provide a short list of approved methods: Hazard and Operability Study (HAZOP), What-If, What-If Checklist, Failure Mode and Effects Analysis (FMEA), Fault Tree Analysis, or an appropriate equivalent methodology.

HAZOP

A key feature of HAZOP is the application of so-called guidewords to deviations from the design or operational intent (see Figure 5-5). Process variables and guidewords used for HAZOP are found in Table 5-2.

There are a few drawbacks to HAZOP. Studies can be tedious, especially if the team spends too much time discussing low-consequence failures. Another shortcoming associated with HAZOP is the tendency to give inadequate attention to failures (causes) that are common to a number of high-consequence deviations.

However, HAZOP is a good technique to use for studying new or existing processes using process flow diagrams (Figure 5-4) or P&IDs. It is flexible and well accepted throughout industry. Excellent software and training for use is readily available.

Table 5-2 HAZOP Deviation Guide[13]

Process Variables	Guide Words						
	No Not None	Less Low Short	More High Long	Part of	As well as Also	Other than	Reverse
Flow	No flow	Low rate Low total	High rate High total	Missing ingredient	Misdirection Impurities	Wrong material	Backflow
Pressure	Open to atmosphere	Low pressure	High pressure				Vacuum
Temperature	Freezing	Low temperature	High temperature				Auto-refrigeration
Level	Empty	Low level	High level	Low interface	High interface		
Agitation	No mixing	Poor mixing	Excessive mixing	Mixing interruption	Foaming		Phase separation
Reaction	No reaction	Slow reaction	Runaway reaction	Partial reaction	Side reaction	Wrong reaction	Decomposition
Time, Procedure	Skipped or missing step	Too short Too little	Too long Too much	Action(s) skipped	Extra action(s) (shortcuts)	Wrong action	Out of order Opposite
Speed	Stopped	Too slow	Too fast	Out of synch.		Web or belt break	Backward

What-If and What-If Checklist

The "What-if" methodology is another easy, flexible PHA technique to use, and is frequently used in combination with HAZOP to study both procedural and equipment hazards. It is especially useful for analyzing existing process systems and involving chemical operators in reviewing procedures.

For a basic study, go through the operating procedures step by step, addressing the what-if questions as the team asks them. As a more organized alternative, record the questions first, then address consequences and causes.

Many facilities use a checklist in combination with the what-if method. Figure 5-8 is an example of such a checklist.

The what-if method can be very effective if the leader structures the study. Otherwise, questions may be asked randomly and lead to superficial results.

Failure Mode and Effects Analysis (FMEA)

FMEA is another easy PHA application. It begins the scenario development process with a study of component failure mode (cause: for example, pump failure).[14] The scenario is next developed by considering the immediate effects (deviation: e.g. high level upstream). Immediate effects are followed by effects or consequences affecting the system (overflow tank, potential contamination of ground water). Using the FMEA

---

1. **Introduction**

   The following checklist may be used to derive "what-if" questions and cover important aspects of a production process operation. The words and phrases in the list should serve to stimulate questions concerning the subject. Action items may be assigned to operations, engineering, or technical functions.

2. **Example**

   The item "Procedures" under "Process Equipment, Facilities, and Procedures" should lead to such questions as:

   ❏ What if we lose power or other utilities during startup or shutdown?

   ❏ What if a critical step is missed during startup or shutdown?

   ❏ What if an emergency shutdown is required during a critical step in the procedure?

   (continued on following page)

Figure 5-8 Checklist for What-If Analysis

Category	Simplified Process Hazards Analysis Checklist Plant_____ Process_____ Subjects to be investigated		Operations	Engrg.	Technical	Date Completed
Storage of raw materials, products, intermedi- ates	Storage tanks Dikes Emergency valves Inspections Procedures Specifications Limitations	Design, separation, inerting Capacity, drainage Remote control, hazardous materials Flash arrestors, relief devices Contamination, prevention, analysis Chemical, physical, quality, stability Temperature, time, quantity				
Material handling	Pumps Ducts Conveyors, mills Procedures Piping	Relief, reverse rotation, identi- fication Explosion relief, fire protection, support Stop devices, coasting, guards Spills, leaks, decontamination Ratings, codes, cross-connec- tions				
Process equipment, facilities, procedures	Procedures Conformance Loss of utilities Vessels Identification Relief devices Review of incidents Inspections, tests Electrical process Operating ranges Ignition sources Compatibility Safety margins	Startup, normal, shutdown, emergency Job audits, shortcuts, suggestions Electrical, heating, coolant, air, inerts, agitation Design, materials, codes, access Vessels, piping, switches, valves Reactors, exchangers, glassware Plant, company, industry Vessels, relief devices, corrosion Area classification, conformance, purging Temp., press., flows, ratios, consequences Peroxides, friction, fouling, compressors, static, valves, heaters Heating media, lubricants, flushes, packing Cooling, contamination				

Figure 5-8 Checklist for What-If Analysis (continued)

method, the study team considers various failure modes for specific equipment components and then evaluates the effects on the system or plant. As when using other PHA methodologies, concentrate on system consequences of interest (major unwanted events such as fire, explosion, and toxic release).

Sometimes additional analysis produces what Johnson calls *criticality estimates* (impacts), *failure mode frequency estimates*, and *detection safeguards* (prevention, protection, and mitigation).

For example, consider a temperature control loop broken down into its component elements: sensor, transmitter, controller, and control valve. Then, for each item, credible failures that have significant consequences are developed. If safeguards are judged inadequate, recommendations are made to remove, protect against, or mitigate the hazard.

Other examples of equipment items and failure modes often considered during an FMEA study are:[15]

- Control valve failure: fails to closed, open, or stuck position. Leaks when closed or massive valve body failure

- Pump failure: stops when running, fails to stop when intended, cannot start when intended, starts when not intended, or cannot deliver desired pressure

- Transmitter failure: gives erroneously high or low reading, fails to respond to process change

- Heat exchanger failure: leaks from process to service side or from service side to process side, leaks to atmosphere, or heat exchange surfaces fouled.

FMEA is a relatively easy application and software is available that includes risk ranking capability. Like HAZOP, it requires detailed process design information. Therefore, FMEA can only be used after the detailed design stage is complete.

Fault Tree Analysis

Fault Tree Analysis (FTA) is a structured PHA technique to examine failure causes and ranking that is also used for incident investigation. The format is a tree structure with the undesired, principle event at the top. Contributing factors are subdivided into events as shown in Figure 5-9. The study in the figure is for a fire or explosion in a furnace used for high temperature steam-cracking of hydrocarbons.

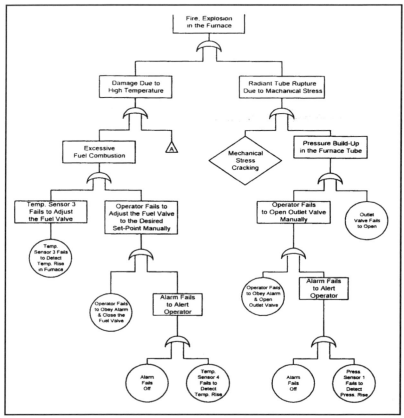

Figure 5-9 Application of Fault Tree Analysis[16]

Fault tree analysis begins with the top event and then moves on to consideration of intermediate events such as high temperature damage in the illustration. Each branch of the tree ends with a basic fault (or cause) that eventually leads to the top event. Relative to safe work conditions, potential equipment failures in this example are

- Temperature sensor fails to detect temperature rise
- Pressure sensor fails to detect pressure rise
- Alarm fails off
- Outlet valve fails to open

Fault Tree Analysis is graphical, well structured, and easily understood. It is a flexible technique applicable to a range of process complexities. Also, this method is intuitive, moving from the general to the specific.

Table 5-3 Advantages and Limitations of PHA techniques[17]

Technique	Advantages	Disadvantages
Checklist	Easy application without much preparation. Relates to most plant situations- material storage and handling, facilities and process equipment, procedures, controls, PPE, fire protection, sampling and waste disposal.	Effectiveness is highly dependent on quality of checklist and attention to detail. Invites superficial hazard analysis without preparation including gathering PSI. May be difficult to revalidate.
What If	Easy application to involve operators evaluating procedures. Effective to use in combination with HAZOP (process flow or P&ID diagrams) or checklist method to prompt "What If—?"	Results may be superficial. What if questions can become random without structuring the PHA study.
HAZOP	Easy application to process blueprints. Well accepted and standardized. Excellent software available.	Time consuming, and sometimes boring. Needs trained team and good process knowledge. Requires updated P&IDs.
FMEA	Easy application. Includes risk ranking and sub-systems analysis.	Time consuming, especially if analyzing non-dangerous failures. Inadequate consideration on common mode or combination of failures.
Fault Tree Analysis	Structured technique. Well used and good to examine failure causes and ranking.	Time consuming. Difficult interpretation- uses complex logic with mathematics. Need current database.

Table 5-3 lists advantages and limitations for the PHA techniques discussed.

Readers need to recognize the flexibility offered by the features of these PHA techniques. Note that the people and their input is more important than the PHA methodology used. Study leaders need to establish an open, creative, problem-solving climate, and sense when the team is "burned-out." More "thinking outside the box" is needed, i.e., emphasizing prevention of initiating events rather than too much focus on existing safeguards. Many companies are now integrating design and safety reviews focused on inherent safety (see Chapter 3: Incident Prevention via Elimination, Reduction, and Moderation) into their process hazards analysis system.

During PHA studies, there is also a tendency for hazard identification to "degenerate" into risk evaluation when the team starts to get tired. Sometimes statements are made such as "This has never happened before (interpretation: 'therefore it will never happen, and let's move on')." Without knowing it, participants may lapse into making risk judgments without data, rather than focusing on the detail of credible hazards and resulting scenarios.

Risk Analysis

An important point about risk determination is illustrated in recounting the author's classroom experience where two Dupont instructors described a potentially catastrophic scenario involving ammonia (toxic and flammable). They provided the class with enough information to estimate the possible consequences of the event. There was general agreement among different groups as to what could be the extent of the disaster. Then the groups were asked to discuss the situation and come to some agreement regarding the probability that this scenario would occur. There was so much disagreement within the groups that the instructors called a halt and announced, "See, you folks have just made the point. It is not hard to agree on what *might* happen (consequences). But when it comes to the likelihood that this thing *will* happen (probability), this argument happens every time we run the exercise."

PHA teams need to perform some kind of risk analysis, however informal, in order to rank action priorities. Common categories to define severity of consequences and probability of occurrence are:[18]

Consequences:
1. Slight to none
2. Low (light injury, small property damage, public nuisance, etc.)
3. High (severe injury, significant property damage, public aroused, etc.)
4. Catastrophic (fatality, major property damage, cease and desist order, etc.)

Probability:
1. Very low (less than 0.0001 experiences per year, or less than 1 per 10,000 or more years)
2. Low (0.01 to 0.0001 exp. per year, or 1 per 100 to 10,000 years)
3. Medium (0.01 exp. to 1 per year, or 1 per 1 to 100 years), and
4. High (1 or more experiences per year)

Table 5-4 Risk-Rating Matrix[19]

	Low Frequency or Likelihood	Medium Frequency or Likelihood	High Frequency or Likelihood
High Consequence	Caution (remedial action may be considered)	Avoid (remedial action required)	Avoid (remedial action required)
Medium Consequence	Safe (remedial action not required)	Caution (remedial action may be considered)	Avoid (remedial action required)
Low Consequence	Safe (remedial action not required)	Safe (remedial action not required)	Caution (remedial action may be considered)

Frequency and consequences may also be described in more qualitative terms (Table 5-4).

The study team should provide as much risk information as possible to high-level decision-maker(s) that are aware of or have been involved in the hazards analysis.

Conclusion

Unsafe work conditions, unsafe work practices, or often a combination of both cause most accidents. Unsafe work conditions are particularly troublesome because it is often the case that they have existed at the workplace for some time. If an accident occurs, OSHA may consider this negligence willful. The OSHA and EPA general duty clauses may then be applied to violations where there is no specific standard.

Further, facilities that neglect unsafe work conditions send the wrong signal to employees. Despite good intentions, the message is "We don't care about safety around here." Good safety programs require regular inspections of all areas with particular attention to high-hazard operations. Hazards are identified, reported, and corrected promptly.

Chemical safety management requires that facilities address all generic hazards (machine guarding and lockout-tagout, electrical, ergonomic, etc.) plus those associated with chemical release and exposure. Facilities handling chemicals must be concerned with spills and splashes, as well as process upsets that could result in a major catastrophe.

Assurance of safe work conditions is a vital part of hazard identification and control. The goal is to acquire hazard information and address harmful deviations from normal via physical inspections (visible hazards) and process hazard studies (hidden hazards). Policies, procedures, and other documentation involving facilities and equipment are needed that meet or exceed regulatory requirements (Chapter 8). This means that employees need site plot plans, piping and instrumentation diagrams, written standard and special operating procedures, material safety data sheets, hazard checklists, air sampling and monitoring devices, plus other sources of safety, health, and environmental information.

Inspections and hazard studies must be planned as to types, frequencies, and participants. It is essential that all facilities, including those with chemicals, have a functioning self-inspection program. Attention must be given to prompt corrective actions, using the Safety Precedence Sequence (Figure 4-6) to determine the most effective approach.

The elements of process safety management have an important place in assuring safe work conditions. Facilities need current process safety information for chemicals, process technology, and equipment and safety systems. Here, process hazards analysis (the methodology appropriate to the hazards involved) is the main tool for identifying and controlling hazards. Distributed control systems and written operating procedures are described to optimize the operator-machine interface. Management of change procedures needs to be in place to control changes that affect process safety information. A healthy mechanical integrity program should be in place, having equipment inspections and tests that ensure safe operation for the life of the process.

Finally, facilities need to understand the basics of risk analysis to assure safe work conditions. Equipment operating for 5 years without failure represents data, not assurance that there will be no failure in the future.

References

1. Adapted from Faisal I. Khan, B.R. Natarajan, and S.A. Abbasi, "Avoid the Domino Effect via Proper Risk Assessment," *Chemical Engineering Progress*, October 2000.

2. www.osha.gov

3. Occupational Safety and Health Reporter, September 15, 1993, OSHA, Department of Labor.

4. Jack E. Daugherty, *Industrial Safety Management, A Practical Approach*, Government Institutes, ABS Group, Rockville, MD, p. 140, 1999. Reprinted with permission.

5. Adapted from *Accident Prevention Manual for Business and Industry, Administration and Programs, 12th Edition,* National Safety Council, 1992, p. 255. Material used with permission.

6. Modified from *Accident Prevention Manual for Business and Industry, Administration and Programs, 12th edition,* National Safety Council, p. 270, figure 11-9.

7. OSHA regulation 29 CFR 1910.119 Appendix B, p. 361, revised as of July 1, 2001.

8. Generated by Hazard Review LEADER™, EQE International, Division of ABS Group, Inc., Rockville, Md. Reprinted with permission.

9. Kevin E. Smith and David K. Whittle, Six Steps to Effectively Update and Revalidate PHAs, *Chemical Engineering Progress,* January, 2001. Reprinted with permission.

10. Ethanol 2000, Bingham Lake, Minnesota. Reprinted with permission.

11. Smith and Whittle.

12. Contact AIChE at www.aiche.org/publcat for more information.

13. Reprinted with permission of the Process Institute, www.jbfa.com.

14. Robert W. Johnson, "Analyze Hazards, Not Just Risks," *Chemical Engineering Progress,* July, 2000, Table 1.

15. Chemical Process Report, Tab 400, page 53, March 1992, Thompson Publishing Group, Washington, D.C. 20006.

16. Hamid R. Kavianian, "Process Safety Management of Potentially Hazardous Chemical and Petroleum Processes," *Professional Safety,* American Society of Safety Engineers, September 1998, Figure 3. Reprinted with permission.

17. Adapted from Anthony E.P. Brown, "Risk Analysis: An Investment in Engineering," *Process Safety Progress,* Vol. 18, No. 2, Summer 1999, Table 1. Material used with permission.

18. *Plant Guidelines for Technical Management of Chemical Process Safety,* CCPS, 1992, Appendix 6B, page 73.

19. Stephen J. Wallace, "Using Quantitative Methods to Evaluate Process Risks and Verify the Effectiveness of PHA Recommendations," *Process Safety Progress* (Vol. 20, No. 1), March 2001, Figure 2. Reprinted with permission.

Safe Work Practices

Introduction

This chapter is the companion to Chapter 5, Safe Work Conditions. The goal of this chapter is to help the chemical safety manager develop policies, practices, and procedures that will ensure that his or her people interact safely with their work environment.

A family of safe work practices (SWPs) is clearly one of the most important assets owned by the facility. Installing and maintaining SWPs takes a good deal of management attention, especially at the beginning of a continuous improvement program. Effective management policy powers this engine of change.

A safe work procedure entails a series of steps to conduct a specific work assignment effectively, efficiently, and safely. Examples of work assignments are process startup and shutdown, tank unloading and loading, vessel charging and discharging, sampling, cleaning, mixing, milling, coating, and drying.

Safety Management Procedures

The safety management triad[1]—a system of people, work environment, and policies/procedures—is a useful way to represent a safety management system (Figure 6-1).

People Factors

Management needs to provide the sponsorship, resources, and decision-making authority to assure safe work practices. They should also take a look at the degree to which the facility culture supports safety. Do people have the training, experience, teamwork, and attitudes to work safely? Keep in mind that personal internal factors such as attitudes, values, and perceptions are extremely hard to shape and control.

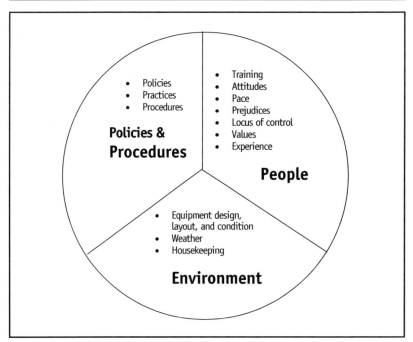

Figure 6-1 The Safety Management Triad[2]

Risk and Reward

One recent book on industrial safety claims that "the number one cause of occupational injuries and illnesses in the U.S. is the workers' inability to detect rising risk factors in their work environment, or, if recognized, the fact that they became concerned too late."[3]

Risk means different things to different people, but we define risk as worker(s) exposing themselves to the chance of an injury or doing something that might cause an unwanted incident. A good example of risk is not wearing safety glasses with side shields (and a face shield) when working with chemicals. We may think to ourselves that we have drained ethanol from a drum into a pail hundreds of times without getting splashed in the face, so what's the big deal now? Or more likely, we may not think of the risk or the hazard at all and just do it without wearing the PPE from habit. Familiarity with a task tends to reduce the perception of associated risk. For example, the more we do an acid sampling procedure without PPE, the less risky it seems.

Another all too common employee perception is that workplace injuries are "fair" or "come with the job." Thus more risk is taken than necessary.

Employee perceptions are difficult to control because the benefits or rewards of at-risk work practices are seen to outweigh the dangers of at-risk behavior. A powerful personal reward for doing work quickly (and often unsafely) is break time. In addition, a pat on the back from the boss for getting product out the door and reinforcement of the importance of productivity at the expense of safety are further rewards that undermine safety. There are also many social rewards for at-risk behavior. Approval from others is a factor. I might work safely when observed, but how do I work without the supervision of the safety coordinator, my supervisor, or the plant manager?

Facility safety coordinators need to help management address the risk and reward factor along with the other people items in Figure 6-1.

Environmental Factors

The environmental factors are associated with safe work conditions, and are the subject of Chapter 5.

Management Policies

Policies and procedures link the people and environment elements in Figure 6-1. Remember, people work *in* the system; therefore management must work *on* the system. Management must do the planning, organizing, implementing, and controlling to direct and drive safe work practices.

Specific goals and objectives must be set. The scope of the change needs to be clearly defined. Resource requirements need to be met. Roles and responsibilities need to be laid out and assigned. Internal communication and coordination are essential to improving safe work practices. Detailed work plans are to be made and milestones for accomplishment set. These are the main considerations, and there are other policy items to put in place to support improving work practices.

The two primary methods for improving safe work practices presented in this chapter are Job Safety Analysis (JSA) and Behavior Based Systems (BBS).

Job Safety Analysis

The job safety analysis (JSA) is a useful tool for documenting tasks that have no written procedure and/or those tasks that are performed only occasionally. The technique involves employees directly in the development of written safe work procedures. It can be implemented at any time, in any part of any facility.

Basic Approach

Figure 6-2 is an example of a job safety analysis applied to the transfer of sulfuric acid from drums to a tote tank. This was a temporary procedure and employees sensed the need to document safe work procedures. Potential hazards and precautions were added later.[5]

The first thing to do with a JSA is to list the steps that are involved in a work procedure in column one. If there is no written procedure, column one can be completed by interviewing the operator.

Next, brainstorm with the operator specific hazards associated with each step. For example, by standing directly over the opening when adding powdered resin to a kettle full of solvent, the operator risks being splashed or exposed to a fire flashback (caused by static ignition in the vapor above the liquid if the static is not sufficiently removed from the resin slabs). The hazards in Figure 6-2 include an acid splash or spill, overfilling the tote tank, having the hose slip loose and dump sulfuric acid on the floor, and generating heat by activating the acid with water.

Finally, discuss precautions associated with each hazard; choose and list them in column three. In Figure 6-2, precautions involve wearing PPE, making sure the hose connections are secure, watching the transfer until complete, replacing the lid, and neutralizing any spilled acid with soda ash. Given the hazards, the worker participates in choosing precautions that will improve the procedure.

How to Use Job Safety Analysis

JSAs bring many benefits. Hazards are found and controlled before the job is performed—thus, before there is a chance of an accident. JSAs are used for employee and supervisor training, accident investigation, facility and equipment inspection, and for uncovering chemical exposure, ergonomic, and other health hazards. JSA is also useful for revising procedures following an incident and for the incident investigation itself. It is a simple, flexible technique for getting right at the hazards involved in any procedure or work situation.

Company X, Y, Z- JOB SAFETY ANALYSIS & JOB INSTRUCTION TRAINING: SULFURIC ACID TOTE TRANSFER PROCEDURE

Job Title:
Worker Title: JSA #001
Department: Page 1 of ___
Supervisor: Date: 6/25/01
Approved by: ☐ New   ☐ Revised

Personal Protective Equipment Required:
Safety glasses with side shields or goggles, facemask, rubber gloves, and chemical suit (specify)

Job Steps	Potential Hazards	Precaution
1. Locate full or partial tote containing sulfuric acid.		
2. Remove small cap and insert acid pump.	Splash or spill	Wear PPE
3. Turn pump on before plugging in cord.	Acid exposure- pump switch is over acid tank	Precautionary step
4. Secure hose from pump to main acid tote.	Hose leak or spill	Make sure hose connection to pump is tight and is secure into main acid tank
5. Pump contents of tote into main acid tote.	Overfill tote / Hose slip loose / Spill	Stay in area and watch transfer
6. When transfer is complete, turn pump off and unplug pump.	Overfill	
7. Remove pump from tote and replace lid.	Water may get in acid. Water mixed with acid will activate the acid.	Replace lid even if tote is not empty.
8. Place pump in designated barrel for storage.		
9. Remove hose from main acid tote and store in designated barrel.		
If you have a spill notify plant and operations mgr.	Spill	Neutralize acid with soda ash. Clean up procedures to be determined

Figure 6-2 Example, Job Safety Analysis[4]

For example, material safety data sheets (MSDS) contain a lot of helpful information relative to hazardous and safe handling of chemicals. However, facilities can apply the MSDS only after knowing what they are trying to do with the material. Alternatively, combining JSA methodology with MSDS information would dramatically advance safe work practices in the chemical industry.

Developing JSA Programs

When developing a JSA program, consider the following factors:

1. Accident frequency—study the facility OSHA log and accident report forms for common types of injuries. Give high priority to procedures and unsafe acts that have a history of (or have a high potential for) causing injuries or occupational illnesses.

2. Job injury and illness severity—give high priority to tasks with injuries involving lost workdays and restricted duty.

3. New jobs or jobs done infrequently—although these jobs may not have any accident history, the associated hazards may not be recognized. JSA is particularly effective as a quick training or retraining tool.

4. High cost per accident—studies show that typically 20% of the accidents cause 80% of direct and indirect costs. Focus initially on accidents/jobs associated with high cost.

5. Other factors—consider factors such as worker age, gender, job experience, working alone, government-regulated jobs, and key/critical jobs in your operation.

Worker Participation

JSA is an effective safety management tool only when workers buy into and actively participate in the program. For example, it is very important to involve the worker directly in documenting the JSA rough draft. It is better to have an imperfect JSA that workers use, than to have the procedure perfectly written and not followed.

By bringing the JSA to the worker, management invites employee involvement and builds ownership into the process of writing safe work practices. Note that many facilities hang or post the completed JSA forms on or near the equipment where they can be used for on-the-job training.

JSAs and Written Procedures

It is common at some plants to have detailed written procedures (standard operating procedures or SOPs) for important jobs, and some choose to use JSA to improve the existing system (as opposed to creating another system). Supervisors and lead operators can apply JSAs to non-routine procedures such as the use of portable solution filters. In many cases, the SOPs are revised to better describe the hazard and safeguard.

Training

JSA has another important function within the chemical safety infrastructure. Called by another name (Job Instruction Training or JIT), JSA can be used directly for training new employees or as refresher training. Using this technique training can be conducted in the classroom or adapted to hands-on, on-the-job training. The written JSA is also used to coach employees observed using at-risk behaviors to do the job more safely.

Behavior Based Systems

Behavior Based Safety (BBS) is more complex than Job Safety Analysis. Based on the psychology of safety, it is a system that is driven by behavioral observation, data, and cultural factors. To produce continuous improvement in safe work practices, organizational values, norms and assumptions about work must be deliberately and gradually shaped. Furthermore, most BBS experts agree that the system cannot be effectively implemented until the organization is "ready," meaning that other elements of the safety and health infrastructure are already in place.

Safety managers are well advised to ask some critical questions of themselves before making BBS a part of their chemical safety management program. After all, people tend to do what is reinforced. Supervisors and managers need to ask, "What reinforcements do I want to avoid?" Once that question is answered, they can ask "What *do* I want to reinforce and how can I make the reinforcers more positive, immediate, and certain?" Use of positive consequences is the most powerful tool for change within behavior-based systems.

Basic Principles of Behavior Based Systems

Many years before behavior modification was considered a science, psychologist B.F. Skinner developed the theory that all behaviors are a function of "antecedents" or activators, and, perhaps to a larger extent, the "consequences" of those behaviors. Aubrey Daniels was one of the first to apply Skinner principles to workplace productivity, quality, and safety and to emphasize the importance of using positive reinforcement in the ABC model.[6]

Figure 6-3 provides the classic antecedent-behavior-consequences (ABC) model associated with BBS.

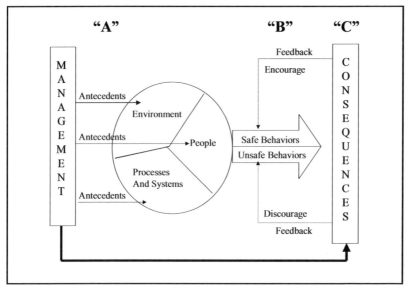

Figure 6-3 Behavioral Safety Model[7]

The elements of the triad from Figure 6-1 interact to form the factors that direct behavior (antecedents) and drive behavior (consequences) in Figure 6-3. Consider the following simple example:

A worker is filling a large solvent storage tank by pumping the solvent methyl ethyl ketone from a rail tanker. The worker is standing in a control room when a blinking signal on the monitor screen reminds him the tank is full. The signal is an antecedent (A), activator, or prompt to shuts off the pump (B, behavior). If the worker does not act, a second antecedent, a high-level alarm, sounds. When the tank is full and the pump shut off, the monitor message reads "fill operation complete." This message is feedback to the worker that the desired consequence (C, fill operation complete) has been accomplished.

Thus the ABC model has worked. The antecedent (blinking signal) has cued the worker to perform the desired behavior and the fill operation complete message has positively reinforced the behavior. Due primarily to the feedback and positive reinforcement, the behavior is more likely to be taken the next time this task is performed.

Consider the consequences of an unsafe behavior. Assume the worker's supervisor was in a rush to get the tanker unloaded. The instructions were "I need this rail car empty for something else right away and I want you to drain it double-time." So the worker thinks a moment and then adds a portable pump to the one already operating. Because there is something wrong with the level alarms (unsafe condition), the worker is told to watch the level indicator carefully. The level indicator is on the tank, far away from the control room, and the worker is distracted. Soon, the tank overflows. The consequences now are negative (clean up the spill) and could have been much worse (fire or explosion if the solvent vapor had found an ignition source). This time the consequences tend to discourage the behavior (turn off the pumps when tank is full). The consequences to the worker would have been even more negative if the supervisor punished or fired him as a result of the incident.

Use of Antecedents, Feedback and Positive Reinforcers

From the behavioral safety model, we see that antecedents, feedback, and consequences are the three factors that influence worker behavior.

Antecedents

Figure 6-3 shows that antecedents are directed to people, the work environment, and work systems. It is easier and more useful to understand antecedents in terms of examples used in the work place:

*   Safety rules and warning signs such as "No Smoking" and "Wear goggles or a face shield with safety glasses when sampling."
*   Written procedures such as found in a Job Safety Analysis.
*   Training sessions covering operations such as chemical reactions, distillation, and mixing/milling.
*   Safety results and activities posted in the break room.
*   Memos, oral and written objective statements, performance appraisals, and other statements of management expectations.

Antecedents can be powerful management tools if used correctly. Unfortunately, they are overused, misunderstood, and often ineffective (especially in the United States).

For example, a memo to a supervisor, expecting him to reduce chemical-related accidents in his department by 50% by the end of the year, is an antecedent. The memo states a management expectation of what the supervisor will do. However, what if the supervisor does not know how to accomplish this objective?

It would be helpful if the memo offered resources such as greater use of engineering controls, updated PPE based on hazard assessments (an OSHA requirement), and specific hazard communication training. Possibly the supervisor will locate these resources, but perhaps the supervisor needs some help from a safety professional. To be effective, antecedents must be accompanied by a clear understanding of the behavior, or what is to be done. The level of detail required depends on the experience and maturity level of the worker.

Behaviors

Behavior is a word that is often misunderstood. It means different things to different people. As one plant manager, new to the BBS concept, said, "Shucks, my people behave. The problem is that they won't work!"

Figure 6-3 indicates behaviors are either safe or unsafe. BBS professionals often teach facilities to speak in terms of behaviors that are safe or *at risk*. For example, following an observation, say to a worker, "You were 80% safe in doing this job, but there were 3 at-risk behaviors." Then reinforce the worker for the safe behaviors and provide the kind of feedback that will cause him or her to change the at-risk behavior.

To be most useful in the continuous improvement process, behaviors need to be positively stated. Use of permits for hot work and confined space entry, wearing PPE when handling chemicals, and correct lifting and body positioning are all important behaviors that are observable and measurable. Give careful attention to documenting a list of important behaviors (called the critical behavior index or CBI) that are generated by employees.

Critical Behavior Index

We want to treat safe and at-risk behaviors as objectively observable and measurable acts, not as deep psychological mysteries. Focus on a few critical things people should do to ensure their safety and health (Figure 6-4).

Observer: _____	Location: _____	
Date: _____	Task: _____	
Time: _____	Number of person's observed: _____	
Weather/Work conditions: _____		
Behavior Type	**Safe Behavior**	**At-Risk Behavior**
Following written procedures		
Use of personal protective equipment (PPE)		
Proper body position		
Eyes focused in direction of movement or task		
Use of permits for lockout-tagout, line-breaking, confined space entry		
House keeping: maintain aisles clear, staging and storage in designated areas		
Cleans up spills and water on floor		

Feedback and comments on back of form.

$$\% \text{ Safe Behaviors} = \frac{\text{Safe Behaviors}}{\text{Total Behaviors}} \times 100 = \text{\_\_\_\_\_} \%$$

Figure 6-4 Sample Behavior Observation Form

The form provided is intended to be comprehensive and is not in actual use. It illustrates typical behaviors that teams of trained workers choose as critical to safe work practices. In actual practice, no more than three to five behaviors should be listed per procedure. Also, the third item should be broken down into a few key behaviors for each of hot work, line-breaking, confined space, and lockout tagout procedures observed.

Use several categories when developing a CBI checklist. Safety researchers Paul Ray and Austen Frey provide such a list from a sampling and observation form:[8]

Housekeeping

1. Equipment is clean.
2. No accumulated dust or dripping oil or grease.
3. Aisles/walkways clear.

4. Containers stored.
5. Eyewash/showers clean and operational.

Personal Protective Equipment

6. Hearing protection (if noise above 85 db).
7. Safety Glasses (with side shields), goggles, and/or chemical face shields.
8. Enclosed, non-skid, electrostatic-dissipating (ESD) shoes.
9. Protective wear like flame resistant coveralls.

Personal Clothing

10. No loose clothing.
11. No jewelry.
12. Hair contained.

Materials Handling

13. Safe lifting and handling of heavy objects such as 55 gallon drums.
14. Use both hands when lifting.
15. Chemical trailers chocked and leveled.
16. Use of dock locks during loading and unloading procedures.
17. Forklift slows and sounds horn at corners and intersections.
18. Stack height per good manufacturing practice guidelines.

Operations

19. Machine guards in place.
20. No hands in equipment point-of-operation.
21. Confined space entry permit displayed and procedures followed.
22. Line breaking and lockout procedures followed.

Feedback and Consequences

The dotted lines from consequences to the behavior arrow in Figure 6-3 are feedback loops. That is, information feedback helps the worker associate the consequence with the behavior, such as in the rail tanker-unloading example above. Feedback is information that originates from the outside and is processed within by the worker. Feedback comes from many sources. It may be information on a bulletin board, data coming from a computer display, or feedback provided by the employee's supervisor. Feedback either encourages or discourages the behavior performed.

Consequences are those events that occur as a result of behavior. Too much attention has been paid to the use of antecedents than in designing and delivering effective and positive consequences.

Positive consequences or positive reinforcement are used to encourage safe behaviors. Negative reinforcement is used to discourage unsafe behaviors. Short hand for positive and negative reinforcement is R+ and R-, respectively.

Professional behaviorists teach that consequences are determined to be positive or negative by the perception of the performer, not according to the intended effect. Also, consequences that are positive, immediate, and certain (shorthand is PIC) are more powerful than other R+. For example, the taste of a cigarette to a smoker is positive, immediate, and certain—thus highly reinforcing.

In a similar fashion, negative reinforcement that is negative, immediate, and certain (shorthand NIC) is stronger than R- that is negative, future, and uncertain. In the smoker example, one of the negative reinforcers is the Surgeon General's message on the label. But the consequence of cancer is in the future and is uncertain—and therefore is a much less effective negative reinforcement than the PIC effect of smoking a cigarette.

Potential Reinforcers

Figure 6-5 lists a variety of things that most people find reinforcing. The list can be used to reinforce actions, behaviors, or procedures that managers and supervisors want to strengthen.

The most important thing to remember when using a list like this is: given the situation, what is reinforcing to the person or group of interest? For example, verbal praise for safety in front of co-workers may be more embarrassing to some people than reinforcing. Often, it will be far more effective to give personal praise in private.

Another key point about providing positive reinforcers is that they should be unexpected. A gift certificate given to an individual following a safety contest is a *reward*, something quite different from a reinforcer. A contest sets up an expectation for a possible reward in advance of behavior or performance. A reinforcer is a consequence of behavior and may be anticipated but not expected.

Implementing Behavioral Based Safety Programs

Conventional BBS programs rely on observations made by volunteer workers, although there are self-observation and feedback systems. Generally, individuals or teams make observations during specific times during the week. The worker being watched knows the observation is coming.

1. Letters of commendation
2. Asking person for advice or opinions
3. Verbal praise
4. Letting the person report his/her results to upper management
5. Increased responsibility
6. Allowing person to make decisions affecting work, organization, strategies or plans
7. Memo to superiors on performance of subordinates with copy to subordinate
8. Passing along compliments from others
9. Choice of tasks
10. Put positive information into personnel folder
11. Remove constant supervision requirement
12. Early start on vacations
13. Represent department at meetings
14. Spruce up work area
15. Time-off
16. Secretarial service
17. Positive comments on performance improvement
18. Exception to company policy or procedure
19. Job transfers
20. Quick follow-up on requests, problems, etc.
21. Name on bulletin board for meeting some goal
22. Training for better job
23. Additional help
24. Talking to person about some outside interest
25. Work on special project
26. Assist supervisor/manager in some special duty
27. Thank you, nod, smile, handshake, or pat on back
28. Personal phone call or note from you
29. Work scheduling
30. Job rotation
31. First choice at extra training and new equipment or tools
32. Talking to person about some anticipated positive reinforcement
33. Listening
34. Promotions
35. Raises
36. Flex-time
37. Bonuses
38. Fringe benefits
39. Parking spaces
40. Car fare or mileage to work
41. Car pool using company van
42. Cup of coffee, donuts, free use of vending machines
43. Gift certificate
44. Plaques, trophies, diplomas
45. Clothing like t-shirts, hats, jackets with special logo or insignias
46. Free lunch, dinner for two
47. Article with special logo or insignia- coffee mug, pen, tie clip, pocket knife, pin
48. Other tangibles of small to large economic value

Figure 6-5 Potential Reinforcers[9]

BBS program leaders must deal up-front with the considerable psychological and social issues associated with observations.

Common Implementation Problems

We indicated above that an attempt to install a BBS is doomed to failure unless the organization is "ready," meaning that important elements of the safety and health infrastructure are already in place.

Six other common errors that companies make when trying to implement a behavioral safety process are:[10]

1. Risk #1: Thinking that observation and participation are the core of behavior-based safety
2. Risk #2: Failing to apply positive reinforcement systematically and effectively
3. Risk #3: Changing only the hourly employees
4. Risk #4: Making behavior-based safety the primary responsibility of the employees
5. Risk #5: Not training managers, supervisors, and hourly employees in the core principles of behavior change technology
6. Risk #6: Trying to fit an activities-based "program" to your organization

Some companies have evaluated their entire safety program to determine their state of readiness to implement BBS. This evaluation process is time-consuming but well worth the effort. The risk factors are real, determined by experience with many user-organizations.

Key Success Factors

Seven factors that are critical to the success of behavior-based processes are:[11]

1. Use of a process blueprint
2. Emphasize communication and buy-in.
3. Demonstrate leadership (both management and labor).
4. Ensure team competence.
5. Use action-oriented training.
6. Use data to ensure continuous improvement.
7. Provide technical resources.

The first four factors are most important in determining BBS success or failure. Because BBS is a significant investment that takes time to implement, the reader is advised to approach top management carefully. Safety, health, and environmental managers will do well to direct some key questions to facility management:

1. When was the last OSHA audit and how well did we do? (Note that audits are more comprehensive than inspections.)
2. When was the last EPA or state pollution control agency audit and how did we stand up to their scrutiny?

Process

Using a process blueprint means we have to be clear about what we are trying to do and the sequence of steps needed to get there successfully. The chemical safety management structure of planning, organizing, implementing, and controlling framework described at the beginning of this chapter is an example of this process blueprint. But we must be flexible in shaping the process to the culture and vice versa. Furthermore, the focus must be on practical change rather than visionary goals and ideas.

The BBS provider will define and help the facility implement the improvement process. Of course, outsiders can only go so far. New BBS users should be prepared for one to two years to see signs that the BBS program is becoming "habit strength."

Behavioral Safety Resources

The reader needs to evaluate several BBS providers and make a careful assessment of the facility needs before making a decision to implement a program. Include the following resources in your evaluation:

1. Marsh Risk Consulting, division of Marsh, Inc., www.marsh.com
2. Behavioral Accident Prevention Plan ("BAPP"), Behavioral Science Technology, Inc. (BST), www.bscitech.com
3. Safety Performance Solutions, www.safetyperformance.com

Application of Safe Work Practice Methodologies

Job safety analysis and behavior-based systems represent a specific tool and a comprehensive system, respectively, for achieving continuous improvement in safe work practices. In fact, job safety analysis fits within the BBS framework of how to improve safe work practices.

This section will demonstrate how JSA and BBS can be applied to three hazardous plant operations that have been the source for many incidents within the chemical industry.

Example 1—Safe Work Practices Involving Hot Work

From the discussion in Chapter 5, recall that most regulatory standards focus on inspections to assure safe work conditions. There are some standards however, that describe specific *procedures* that must be followed. JSA or BBS could be applied to these procedures.

For example, hot work (29 CFR 1910.252 (a)) requires the following precautions prior to cutting, welding, and brazing:

1. Moving objects considered fire hazards
2. Guarding objects that cannot be moved
3. Covering cracks or holes in walls, open doorways and open or broken windows
4. Placing fire extinguishers in the immediate area
5. Recognizing conditions requiring a fire watch person
6. Wearing PPE

In addition, the hot work supervisor is to inspect the area and authorize cutting or welding, preferably in the form of a written permit.

Figure 6-6 is a typical open flame and spark hazard permit.

The permit represents a 14-step checklist that meets or exceeds the OSHA standard. Job Safety Analysis could easily be applied to this procedure. As an alternative, the items could be converted into a critical behavior index for cutting and welding.

Consider the six precautions above or the 14 items from Figure 6-6 as critical procedures and behaviors to reinforce using Job Safety Analysis and/or behavioral methods.

Example 2—Confined Space Entry

The OSHA confined space entry standard, 29 CFR 1910.146, is a comprehensive, detailed, and often-overlooked equipment and procedural standard. Its requirements are especially important to facilities storing or processing chemicals because of the breathing hazards the materials can cause. In addition, procedures and natural processes such as painting,

---

OPEN FLAME AND SPARK HAZARD PERMIT

Assigned to (employee or contractor)_____

For (type of work)_____

On (specify equipment)_____

At (building or location)_____

Permit valid from ___(month/day/year)   To: _____(month/day/year)

Time_____ (AM/PM)   Time_____ (AM/PM)

Item (To be completed by authorizer of permit) Check all items	Not Applicable	Complete
1. Remove all flammable and combustible liquids from work area		
2. Monitor area with LFL (lower flammable limit) analyzer		
3. Inert vessel(s). Requires continuous LFL monitoring of work area		
4. Relocate all combustibles 35 feet from work area or shield from hazards with flameproof covers		
5. Protect surrounding equipment with flameproof covers		
6. Protect sprinkler heads, heat activating devices, and flame or smoke detectors		
7. Close all fire doors		
8. Cover or shield all duct, wall, and floor openings with flameproof covers		
9. Fill traps with water		
10. Monitor sewers, trenches, manholes, floor drains, elevator pits, sumps and low areas with a LFL analyzer		
11. Cover floor drains, trenches, manholes and sewers		
12. Provide ___ fire watcher(s), trained in use of portable fire extinguishers and facility emergency procedures. Emergency phone no. _____		
13. Provide the following quantity of fire extinguishers: (__ Water) (__CO$_2$) (__Dry chemical) (__Light water)		
14. Review emergency procedures		

Special requirements

_____

Authorizer	Signature and department number	Phone or pager	Date
Firewatcher's signature	Signature and department number	Phone or pager	Date
Signature- hot work worker	Signature and department number	Phone or pager	Date
Other signatures required by facility	Signature and department number	Phone or pager	Date
Designated company representative	Signature and department number	Phone or pager	Date
Contractor designated representative	Signature and department number	Phone or pager	Date

Figure 6-6 Open Flame and Spark Hazard Permit

welding, solvent cleaning, inerting, fumigating, decaying plants and animals, and rusting can cause lack of oxygen and/or toxic materials in the air. Confined spaces are large enough for workers to enter and to work inside them. However, these spaces are not designed to be occupied continuously. There are also hazards associated with entering and leaving confined spaces. Hazardous breathing conditions and other lethal conditions are often unseen to the worker.

It is useful to define confined spaces in terms of examples. Tanks and vessels, large pipes, pits, trenches, and other enclosures all may be confined spaces if exposure to hazards is required during entry, and the space is not designed for continuous occupancy.

Safe confined-space work procedures are especially important for two reasons. First, injuries are much more likely to be severe, even to the point of death. Also, it is frequently difficult to rescue an injured worker from a confined space. Many of the confined space fatalities that occur each year are would-be rescuers.

OSHA requires employers to develop a program to prevent confined space injury or death that includes a permit system that documents all safe work procedures before and during the entry. The permit also serves as a checklist for the entry team to follow.

Figure 6-7 is a confined space permit provided in OSHA's standard 29 CFR 1910.146 (Appendix D-1).

Again, JSA and BBS could be applied to procedures derived from the permit.

Confined space accidents often result in fatalities when workers are overcome by an oxygen-starved atmosphere or by toxic vapors. Consider the following safeguards or the 10 permit items as safe procedures, behaviors, or safe work practices:[12]

- The vessel is clean, has been washed, and the liquid analyzed.
- I (entrant) personally review each line to and from the vessel to make sure that it is disconnected and blinded.
- I personally lock out related moving equipment such as an agitator motor.
- Continuous monitoring equipment is in place to measure the amount of oxygen and check for combustibles.

```
Confined Space Entry Permit                        Date and Time Expires:
Date & Time Issued:                                Job Supervisor
Job site/Space I.D.:                               Work to be performed:
Equipment to be worked on:

stand-by personnel

1. Atmospheric Checks:  Time                       8. Entry, standby, and back up persons:  Yes  No
                        Oxygen _____ %               Successfully completed required      ( )  ( )
                        Explosive _____ % L.F.L.     training?
                        Toxic _____ PPM              Is it current?              N/A  Yes  No
                                                                                  ( )  ( )  ( )
2. Tester's signature
3. Source isolation (No Entry):  N/A  Yes  No      9. Equipment:
   Pumps or lines blinded,       ( )  ( )  ( )        Direct reading gas monitor -      ( )  ( )
   disconnected, or blocked                           tested
4. Ventilation Modification:     N/A  Yes  No         Safety harnesses and lifelines    ( )  ( )
   Mechanical                    ( )  ( )  ( )         for entry and standby persons     ( )  ( )
   Natural Ventilation only      ( )  ( )  ( )         Hoisting equipment                ( )  ( )
5. Atmospheric check after                            Powered communications            ( )  ( )
   isolation and ventilation:                         SCBA's for entry and standby      ( )  ( )
   Oxygen _____ %          > 19.5 %                   persons
   Explosive _____ % L.F.L. < 10 %                   Protective Clothing               ( )  ( )
   Toxic _____ PPM          < 10 PPM H2S             All electric equipment listed     ( )  ( )
   Time _____                                        Class I, Division I, Group D
   Testers signature                                  and Non-sparking tools

                                                   10. Periodic atmospheric tests:
6. Communication procedures:                           Oxygen ___ % Time ___  Oxygen ___ % Time ___
                                                       Oxygen ___ % Time ___  Oxygen ___ % Time ___
                                                       Explosive ___ % Time ___ Explosive ___ % Time ___
7. Rescue procedures:                                  Explosive ___ % Time ___ Explosive ___ % Time ___
                                                       Toxic ___ Time ___  Toxic ___ Time ___
                                                       Toxic ___ Time ___  Toxic ___ Time ___

We have reviewed the work authorized by this permit and the information contained here-in. Written
instructions and safety procedures have been received and are understood. Entry cannot be approved if any
squares are marked in the "No" column. This permit is not valid unless all appropriate items are completed.
Permit Prepared By: (Supervisor)
Approved By: (Unit Supervisor)
Reviewed By (Cs Operations Personnel): _____      _____
                                         (printed name)                  (signature)

This permit to be kept at job site. Return job site copy to Safety Office following job completion.
Copies: White Original (Safety Office)  Yellow (Unit Supervisor)  Hard(Job site)
```

Figure 6-7 Example Confined Space Permit[13]

- Air movers are in place to move fresh air through the vessel and keep it cool.
- A trained worker is posted at the entry to sound an alarm in case of problems.
- I carry a five-minute air escape pack in case of emergency.

Example 3—Contractor Safety

Regulatory requirements for contractors are found in OSHA's 29 CFR 1926 construction code. For years, the injury and illness rate in the construction industry has been the highest among all industrial and commercial classifications. Because of the number of contractor and subcontractor firms that are often on-site at one time, it is difficult for OSHA to determine who is actually responsible for safety violations. Furthermore, "general contractors and subcontractors often simply failed to coordinate their responsibilities for worker safety. In response, OSHA began citing host employers for safety violations committed by independent contractors and their employees."[14]

In another development, many of the chemical process industry catastrophes that have occurred in past years have involved contractors. Authors of the OSHA process safety management standard included representatives from industry. They paid particular attention to this area, requiring the employer to work with the contractor to assure that maintenance and repair, turnaround, major renovation, or specialty work on or next to covered processes is done safely. OSHA made the site employer responsible for the following:

- Informing contractors of potential fire, explosion, or toxic release hazards associated with the contractor's *work procedures* and the process

- Selecting contractors based on an evaluation of the contractor's safety performance and programs

- Developing and implementing *safe work practices* to control the entrance and presence of contractors

- Explaining the provisions of the facility emergency plan

- Periodically evaluating contractor performance

- Maintaining a contract employee injury and illness log

The contract employer is required to:

- Assure that contract employees are instructed in the known potential fire, explosion, or toxic release hazards and the applicable provisions of the facility emergency response plan

- Document that contract employees have received and understood required training that includes employee name, date of training, and means used to verify that the employee understood the training

CONTRACTOR INFORMATION FORM

(TO BE COMPLETED BY SUPERVISOR/ENGINEER REQUESTING SERVICE)
COMPLETE TOP PORTION AND RETURN TO PLANT SAFETY

CONTRACT COMPANY/ADDRESS:_____

NAME OF COMPANY REP: _____ PHONE #:_____

ON SITE (LIST DATES): _____ FROM: _____ TO: _____

WORKING HOURS (LIST TIMES)   FROM:_____ TO: _____

AREA WORKING:_____ BLDG#:_____ IF PSM AREA:(X)

WORK INVOLVED (BE SPECIFIC):

_____

_____

IS VESSEL ENTRY INVOLVED? (YES/NO)_____

IF YES, LIST VESSEL/CONFINED SPACES:_____

IS HOT WORK INVOLVED? (YES/NO)_____

IS LOCKOUT INVOLVED? (YES/NO)_____   MULTIPLE:____ SINGLE:_____

LIST OTHER REQUIRED SAFE WORK PRACTICES:_____

IS NOISE OVER 85 DBA? (YES/NO)_____   IS HEAT OVER 80 F? (YES/NO)_____

Place an 'X' next to known hazards: a '?' if unknown

Carbon monoxide		Bleach		Paint thinner or solvents	
Lack of oxygen for breathing		Sulfuric or other strong acids		Methanol or Ethanol	
Toxic chemicals (see MSDS)		Flammable vapor		Sodium Hydroxide, caustics, and other strong bases	
Ammonia		Ignitable powder or dust		Hydrogen Sulfide	
Chlorine		Asbestos		Lead	

LIST OTHER CHEMICALS BEING BROUGHT ON SITE:

**TO BE COMPLETED BY SAFETY & RETURNED TO
SUPERVISOR/ENGINEER REQUESTING SERVICE**

AFTER REVIEW OF DOCUMENTATION SENT TO PLANT SAFETY, THIS COMPANY HAS
BEEN CLASSIFIED AS: QUALIFIED ( ) NOT QUALIFIED ( )

PLANT SAFETY SUPERVISOR:_____ DATE:_____

Figure 6-8 Contractor Information Form

- Assure that contract employees follow plant safety rules, including *safe work practices*
- Advise employees of any unique hazards presented by the contract employee's work

Consider this set of requirements as an extension of the general duty clause. Employers must "furnish to each of his employees employment and a place of employment which is free from recognized hazards that are causing or are likely to cause death or serious physical harm to his employees." This includes contractor employees also, especially since the host employer has primary control over the work site. Contractors must follow safe work practices determined by the employer that are appropriate to the project. This element of the PSM regulation has gone a long way to improve construction safety in the chemical and petroleum industries.

The host employer must evaluate the contractor's performance on-site. Use a form like the one illustrated in Figure 6-8 to define the work and list the potentially hazardous work procedures that are involved in the project.

The information form defines operations where critical procedures must be performed. This is another situation where job safety analysis and behavioral observations may be applied to assure safe work practices.

Conclusions

Chemical safety managers must look to safe work practices to control physical and chemical hazards. Chemical facilities have all or most hazards found in other industries, plus those associated with ammonia, chlorine, and the other halogens; isocyanates; corrosive acids and bases; and highly flammable materials such as hydrogen, alcohols, and ketones. The message to readers is to establish a more rigorous safety and health infrastructure than found elsewhere. The nuclear industry is a good example of an industry where policies and practices are more stringent than those in other industries.

Facility management is responsible to affect policies and procedures—the key element in a safety management system (Figure 6-1). Effective and safe operating and maintenance procedures are needed for line-breaking, addition of static-producing materials into flammable solvents, sampling hazardous materials, hot work, confined space entry, and other routine and special hazardous work.

Chemical safety managers must also be sure to note the cultural atmosphere of the workplace. Internal factors such as attitudes are difficult to observe and influence. A positive attitude toward safety is largely developed through employee involvement. A strong chemical safety management system depends on a policy that sets high expectations for safe work practices and minimizes at-risk behavior.

Readers will do well to apply (in more or less step-wise fashion) two methods to improve safe work practices:

1. Install job safety analysis (JSA) because it is a quick and effective way to document procedures, identify associated hazards, and add appropriate precautions.
2. Install some form of a behavior-based system (BBS) to address the issue of at-risk behavior on a plant-wide scale. BBS is only effective when a significant part of the safety and health support system is in place.

Finally, apply the continuous improvement process for safe work practices to employees and contractors. Show customers and suppliers how to improve safe work practices using your materials and products. Focus on tasks and hazards that have been a primary source of chemical related incidents. These include toxic chemical releases, hot-work procedures that if mishandled can lead to fires and explosions, confined space entry, and the others mentioned above. By carefully applying the concepts presented in this chapter, you will be well on your way to establishing safe work practices.

References

1. Greg A. Barrett, "Management's Impact of Behavioral Safety," *Professional Safety,* American Society of Safety Engineers, March 2000.
2. Greg A. Barrett, "Management's Impact of Behavioral Safety," *Professional Safety,* American Society of Safety Engineers, March 2000, p. 27, Figure 1. Reprinted with permission.
3. Jack E. Daugherty, *Industrial Safety Management, A Practical Approach,* Government Institutes, Rockville, Md., 1998.
4. Ethanol 2000, Bingham Lake, Minnesota. Reprinted with permission.
5. Ethanol 2000, Bingham Lake, Minnesota.
6. Aubrey C. Daniels, *Performance Management: Improving Quality and Productivity Through Positive Reinforcement,* Tucker, GA, Performance Management Publications, 1989.
7. Stephen H. Reynolds, "Back to the Future: The Importance of Learning the ABCs of Behavioral Safety," *Professional Safety,* February 1998 (ASSE), Fig. 1, p. 24. Reprinted with permission.

8. Paul S. Ray and Austin Frey, "Validation of the Behavioral Safety Index," *Professional Safety,* American Society of Safety Engineers, July 1999, p. 26.

9. Aubrey C. Daniels, *Performance Management: Improving Quality and Productivity Through Positive Reinforcement, 2nd Edition,* Performance Management Publications, 1984. Reprinted with permission.

10. Jerry Pounds, "High Risk Safety-The Six Biggest Mistakes in Implementing a Behavior-based Process," *Performance Management Magazine,* Aubrey Daniels & Associates, Inc., Tucker Georgia, Volume 15, Number 4, fall, 1997.

11. John H. Hidley, "Critical Success Factors for Behavior-Based Safety," *Professional Safety,* American Society of Safety Engineers, July, 1998, p. 30-34.

12. Bob Brown, "Confined Spaces: Identify the Hazards Involved," *Professional Safety,* Sept. 1999, p. 58.

13. 29 CFR 1910.146 Appendix D-1

14. Daniel J. Nelson, "Who Is Responsible for Safety?" *Professional Safety,* ASSE, April 2001, p. 21.

Accountability and Performance Measures

Introduction

In order to establish an effective management system, facility management must accomplish three critical objectives:

1. Position chemical safety management as an important management function relative to other business priorities.
2. Set the criteria and management expectations for safe plant operations.
3. Install and reinforce measurable goals to use as a basis for continuous improvement and decision-making. This objective includes implementing an annual safety and health plan.

Too often and for many reasons, operational management leaves the responsibility for safety to employees or the safety manager. Safety is a line management responsibility.

In a chemical safety management system, we cannot talk about accountability without measurement. Interestingly, *valid measurement of safety system effectiveness* continues to be one of the biggest problems facing the manufacturing industry today. Noted behavioral safety expert, Dan Petersen, explains it this way: "To hold someone accountable, management must know whether s/he is performing well. Thus his or her performance *must be measured.*"[1] Another way of looking at this is to say, "Anything worth doing is worth measuring."

Figure 7-1 illustrates upstream (before the accident) activities that control safety outcomes or results of non-conformance.

For chemical safety management programs, add unwanted events such as fires, explosions, chemical spills, and accidental/routine air and water releases to the unwanted results and consequences. Note the important difference between upstream and results (downstream) measures. Tracking accident data (also called "outcomes of nonconformance") is an important practice, but is not very helpful as a means to improve a safety

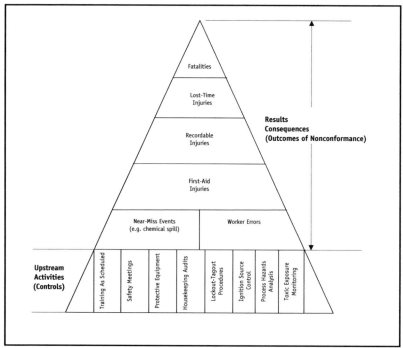

Figure 7-1 Safety Performance Pyramid[2]

record. Therefore, we want to emphasize upstream control measures related to management systems that are built to prevent incidents and accidents.

This chapter examines the components of accountability and how to install safety and health measurements. We will provide examples and tools that can be applied directly to a facility chemical safety management system.

Annual Safety and Health Plan

In the author's experience many plants overlook the benefits of having an annual safety and health plan. Typical example objectives read:

1. Complete PPE hazard assessments for ethanol, acid, and caustic sampling in the operations department
2. Batch chemical department supervisors to complete JSAs for all procedures where operators are exposed directly to hazardous materials
3. Site manager to participate in at least 8 of 12 monthly plant inspections

4. Laboratory safety coordinator to produce a three-ring binder of glove type to be used for each chemical handled

Most safety and health programs need an effective communication plan. Provide access to chemical safety information (include incidents and investigations) for employees, company management, and outside stakeholders. Outsiders include the neighboring community, the public at large, and government regulatory bodies. When it comes to communication media, there are many options. Consider the needs of your audience before expending resources.

There are many benefits to the facility, once a communication plan is installed. These include greater awareness and interest, more stakeholder buy-in, better employee performance, decision making based on better information, and increased public and regulatory confidence.

Once safety and health is made a part of annual business planning, there is a basis for continually improving the safety management system.

Experts in the field of safety accountability and performance measurement have different formulas for achieving continuous improvement, but may take the basic behavioral approach described in Chapter 6.

Management Systems and Accountability

Without accountability, there is no management system. We must be concerned with designing a management system that will set the criteria and expectations for safe plant operations. This means:

1. Setting policy, assigning responsibility, and defining resource requirements

2. Describing safety and health procedures to use as a basis for measurement and feedback

It is useful to look at a typical plant safety and health manual as a framework for establishing a safety management system (Figure 7-2).

Each section listed in the table of contents has a similar but specific management system. Plant programs must be laid out in detail to establish accountability and measure conformance to management expectations. When designing your management system, describe methods for assigning responsibility, determining resource requirements, and establishing feedback systems and audits.

I. **General Policy**
   A. Safety Management System
   B. Line Accountability
   C. Regulatory Requirements
   D. Safety Review and Improvement Process
   E. Employee Involvement And Communications
II. **Accident and Incident Prevention Program**
   A. Emergency Response
   B. Medical and First Aid Provisions
   C. Accident, Incident and Potential Hazard Investigation Procedures
   D. OSHA Recordkeeping Program
III. **Chemical Safety Information**
   A. Chemical Inventory and MSDS Information
   B. Personal Protective Equipment Program
   C. Hazard Communication Program
      1. Hazard Warning Signs
      2. Container Labeling Requirements
IV. **Process Technology and Equipment Information**
   A. Process Technology Information
   B. Process Equipment Information
V. **Operational and Maintenance Policies and Procedures**
   A. Hazard Identification and Control Program
      1. Health Protection and Industrial Hygiene Controls
      2. Safe Work Conditions Program
      3. Housekeeping
   B. Safe Work Practices and Operating Procedures
   C. Mechanical Integrity Program
   D. Maintenance Procedures
VI. **Training, Learning and Growth Programs**
   A. Goals and Objectives
   B. Employee Development Plans
   C. Course Outlines
   D. Training Documentation
VII. **Contractor Safety Program**
   A. Employer Responsibilities
   B. Contract Employer Responsibilities
   C. Contractor Information Forms
VIII. **Management of Change**
   A. Procedures
   B. Change Authorization Forms
IX. **Pre-start Safety Reviews**
   A. Policy
   B. Pre-Start Meeting Minutes

(continued on following page)

Figure 7-2 Plant Safety and Health Manual Table of Contents

> **X. Fire Prevention and Response Program**
>   A. Fire Protection and Prevention
>   B. Portable Fire Extinguishers
>   C. Fixed Fire Protection System
>   D. Fire Detection and Alarms
>   E. Fire and Emergency Exits
> **XI. Site and Transportation Security**
> **XII. Emergency Planning and Response**
>   A. Emergency Action Plan
>   B. Facility Maps and Layouts
>   C. Emergency Evacuations and Drills
> **XIII. Internal and External Audits**
>   A. Policy
>   B. Internal Audit Reports
>   C. External Audit Reports

Figure 7-2 Plant Safety and Health Manual Table of Contents (continued)

It is also important to draft an outline of the basic regulatory requirements (covered here under general policy) to demonstrate that the plant meets or exceeds the regulations. Review 29 CFR 1910 to make sure the safety program complies to OSHA subparts:

H—Hazardous materials

I—Personal protective equipment

J—General Environmental Controls

K—Medical and First Aid

L—Fire protection

N—Materials handling and storage

Q—Welding, cutting and brazing

S—Electrical

Z—Toxic and hazardous substances

OSHA, EPA, and DOT regulatory requirements are discussed in Chapter 8. Use the activities directed to these requirements to establish an upstream baseline level of safety performance.

Components of Accountability

The Center for Chemical Process Safety outlines the following components of accountability that apply to a plant system for chemical process safety:[3]

- Continuity of Operations
- Continuity of Systems

- Continuity of Organization
- Quality Process
- Control of Exceptions
- Alternative Methods
- Management Accessibility
- Communications
- Company Expectations

Here the components will be applied more broadly to include all aspects of chemical safety management.

Continuity of Operations

Continuity of operations means integrating chemical safety with production goals. Safety professionals would like to see safety be first among the equals of safety, productivity, and quality. Line and top managers must make tough decisions to assure continuity of all operations. This includes:

- Manufacturing processes such as coating, drying, distillation, evaporation and dehydration, along with associated safety systems.
- Internal material flow and storage
- Laboratories, pilot plants, warehouse and distribution centers
- Pipelines
- Water and waste treatment facilities
- Power plants, utility and support systems

Responsibility for maintaining continuity of operations often requires management to resolve competing priorities. For example, a small operational problem on a distillation column may eventually mean that the process must be shut down for unplanned maintenance. Operations managers need to make the decision—whether to do it immediately or put off corrective action (which may compromise safety).

The essential element to assure continuity of operations is planning. Facility, engineering, and design people should work together to plan process and equipment features that will minimize shutdown and maximize safety (Chapter 3). For example, plan for[4]

- Additional capacity, particularly in real or anticipated process bottlenecks
- Parallel or "multi-train" process components instead of a single stream

- Ability to shut down sections of the plant independently
- Just-in-time provisions for spare parts

Unplanned shutdowns are counterproductive. Plan semi and annual shutdowns carefully.

Continuity of Systems

It is easy to place continuity of operations ahead of maintaining management systems. Greater attention to management systems would benefit organizations beyond the chemical industry.

This accountability component involves adequately resourcing and funding all parts of the chemical safety system. For example, a process safety coordinator must be in place and PHA teams must be given the time and resources to conduct effective hazard studies. One of the most critical systems for chemical facilities is the gathering and maintenance of process safety information (PSI). Note that corporate and plant management need to lead and follow up on continuity of systems efforts.

Continuity of Organization

Sufficient people-power and monies are needed to assure continuity of the chemical safety organization. Frequently, safety is assigned to an established function such as human resources, plant engineering, or operations without additional resources to do the work. Small and large organizations often have a particular problem when staffing and funding the chemical process safety function because they underestimate its highly operational nature. Process safety cannot come from the outside. It is a critical internal component of manufacturing, process engineering, and maintenance.

Successful organizations staff and fund safety in a creative and effective manner. For example, a 3M facility assigns coordination of process safety activities to experienced process engineers and rotates the job every 2-3 years. Funding for items such as PHA studies and employee training comes out of individual plant product support budgets.

Creating a corporate or regional staff EHS function is another example of building and maintaining safety organization continuity. The staff person is responsible to coordinate activities such as multi-plant PHA studies and to lead audits driving compliance to the PSM regulation. Consulting assistance is sometimes needed, but the organization should be developed from within.

Place a check mark next to each activity that top management at your company does.

___ Issues a written injury and illness prevention policy.

___ Wears appropriate safety gear while touring plant.

___ Discusses safety with employees during periodic plant tours.

___ Is familiar with details of safety program and safety rules.

___ Presents safety awards to employees.

___ Participates as a student, in some safety training programs (e.g., first-aid, CPR, and fire extinguishers).

___ Occasionally attends employee safety meetings as an observer.

___ Keeps informed of leading causes of accidents.

___ Receives copies of supervisors' accident investigation reports.

___ Interviews plant (or department) managers when one of their employees has a lost-time accident.

___ Attends meetings with insurance company safety consultants.

___ Receives copy of insurance company safety reports.

___ Receives copy of safety committee minutes.

___ Reviews regular reports on safety achievements.

___ Ensures that safety is an agenda item at staff and department meetings.

___ Receives insurance companies "Statement of Losses."

___ **Total number of check marks**

Figure 7-3 Top-Management Commitment Checklist[5]

Management should establish a safety budget separate from other functions and forecast key priorities and expenditures each year. In addition to supporting safety, top managers should commit some of their time to safety. Figure 7-3 is a checklist provided by state OSHA to help managers demonstrate personal commitment to safety.

Give your plant manager or department head a copy of Figure 7-3 and encourage them to develop their own checklist.

Quality Process

The essence of this accountability component is the integration of quality and safety. Only a safe process will produce a quality product at low cost. There are many parallels between safety and quality.

Quality-related projects are usually justified based on financial considerations alone. The financial consequences of decisions affecting safety are less direct. However, the philosophies of quality and safety management are very similar. Quality thinking can readily be extended to safety.

To help fold quality and safety together, monitor items such as: percentage process "up-time," percentage of planned versus total shutdowns, monthly product quality measures, and hours worked without injury.

Control of Exceptions

Management of change is the primary system in Figure 7-2 that controls exceptions. Every plant needs an MOC system, regardless of regulated chemical processes that they may have. Changes other than replacement-in-kind that may impact safety need to be reviewed such that the safety systems are still consistent with the needs of the process (see required process safety information). An example of such a change is deactivating a vacuum control or shutting a valve on a vessel normally operating under pressure. Frequently, a tank, used mainly under pressure, will collapse suddenly if the vacuum exceeds its low-pressure limit. Review Figures 5-6 and 5-7 for examples of management of change policy.

Other exceptions made to processes and equipment may be as a result of technical or financial considerations. The most common changes that require safety controls are changes in product demand or feed supply, equipment repair and replacement, and changes in operating conditions.[6] All of these changes need to go through a rigorous review system to evaluate their impact on safety. Install exception control systems early in the facility's life cycle, beginning at the design phase and continuing with effective MOC procedures.

Alternative Methods

Facilities need to have accountability and measurements for alternative management methods. This is especially true if monitoring shows safety performance to be low, flat, or declining. Moving from reliance on Job Safety Analysis to the use of behavioral methods for a procedure like raw material charging is a good example of using alternative methods to improve safety. Carefully consider availability of facility and corporate resources such as materials research and development, engineering systems, technology development, and outside resources. Management capability and experience to implement an alternative project is also needed here.

Management Accessibility

Top organizational management must set a policy that supports plant operations. This means executives need to be aware of facility goals and issues and are accessible when advice or a decision is needed. In addition, plant and staff management needs to be available to assure decisions affecting safety are made quickly. Most organizations put some kind of on-call system in place to assure continuous decision-making coverage. Make sure the culture allows deviation from the chain-of-command where necessary. For example, the plant or technical manager may need to contact an operator directly to approve or make a change affecting safety, following up later with the operations supervisor.

Involve employees in setting up the system. Appoint a process safety coordinator/manager with authority to review changes and make decisions. Provide access to senior management where necessary.

Communications

One all-too-common mistake is to do the safety work and not tell anyone about it. There are many direct and indirect benefits from a strong safety communications effort. These include better personnel and process safety performance, greater employee awareness and participation, better training, information-based decision-making, and increased public and regulatory confidence.[7]

The key component to good safety communications is chemical and process safety information that moves freely in all directions and to all levels. It is important to share this information with management as well as employees. Develop and communicate a file of "lessons learned" from incident data internally and with other facilities in the business system.

Keep operating and maintenance procedures up to date, and use methods such as job safety analysis to identify and control associated hazards. These procedures facilitate communication to affected employees and demonstrate management's ongoing involvement. Use a variety of communication forms: newsletters, memos, posters, bulletin board items, and small group and plant-wide presentations. To get the most out of the communication system, pay particular attention to the intended audience.

The American Chemistry Council's Responsible Care initiative (discussed in Chapter 11) encourages education of all employees about the organization's safety, health, and environmental program. The Responsible Care codes of management practice outline a basic approach to openly communicate with the local community regarding relevant EHS issues.

Company Expectations

Discuss the vision and issues for chemical safety management before finalizing written policy. Set achievable, observable, and measurable objectives to prevent unwanted incidents, unplanned shutdowns, and hazardous impacts outside the plant. Consider important factors such as legal requirements, the financial impact of accidents, and public perception.

Example

Ethanol 2000, an ethanol-producing plant in Minnesota, communicates the following values that guide the operation:[8]

1. All injuries and work related illnesses are preventable.
2. Management is responsible for providing a safe work environment. Safety and health is a line-management function, practiced on a day-to-day routine basis.
3. Employees are responsible for working safely. Safety and health is an inherent job responsibility. Safety responsibility and accountability must be achieved in the same manner as for production, quality, cost, and personnel relations.
4. Safety training is essential. To carry out their safety responsibilities, employees, supervisors, and managers must be properly trained. We require attention to content and context of message, quality of instruction, and follow-up to determine trainee understanding.
5. Safety is an integral part of the business plan. Therefore, it will be a factor in our business decisions. Safety is as manageable as anything else we do in the plant.

The company emphasizes upstream controls over safety results such as OSHA recordables and lost-time injuries/illnesses. They also involve employee action teams directed to education, training, health and environmental programs, rules, safe work practices, inspections, audits, fire and emergency preparedness, accident and incident investigation, and housekeeping.

For continuous and long term improvement:

1. Each employee must complete all procedures and tasks according to standards set by management.
2. Management holds each employee accountable to meet the standards. This is the function of the performance appraisal and employee development plan.

Supervisors are expected to assure that each employee understands what is expected of them. This means measuring and enforcing performance and taking disciplinary actions where necessary. Outcomes of activities at the base of Figure 7-1 are more likely to occur than outcomes towards the top. Do not attempt to work on all systems simultaneously—set priorities. Line management, not the safety department, must show employees how to perform assigned tasks. Line management monitors activities and takes corrective action to see that instructions are followed.

Installing Performance Measures

If safety is truly to become the first among the equals of safety, quality, and productivity, we must measure chemical safety in much the same way as other important areas of accountability. Having accountability in place, the goal is to move on to implement an effective measurement system. Our purpose is to move from simply trying to *manage* traditional safety results to a system of upstream controls that will *improve* safety results (Figure 7-1).

Consider six simple steps to implement an effective measurement system[9]

1. Establish standards.
2. Set priorities.
3. Achieve employee agreement.
4. Establish accountability.
5. Implement an effective communication plan.
6. Measure conformance to instruction.

Establishing Standards

The purpose of this step is to outline and establish activities within a framework of policy like the Safety Management System example above. Work activities must reflect the values shared by management and workers if safety is to become part of the company culture. The idea is to put a set of policies in place that meet or exceed federal, state, and local regulatory requirements (Figure 8-11). Effort spent to establish specific facility standards will pay off in the final step—developing a meaningful measurement system. For example, state that "This facility will comply with all chemical regulatory requirements, including hazard communication, process safety management, and hazardous waste operations and emergency response." Then provide the individual categories of the regulation, for example detailing the elements of process safety management. Helpful performance indicators could be related to safety systems and devices installed, PHAs conducted, MOCs approved, training sessions related to safe operating procedures, etc.

Setting Priorities

This step involves deciding what is most important and how to measure it. The first step is to prioritize areas where the greatest risk reduction can be achieved. A guideline for this process is found in OSHA's PSM standard which requires a "determination and documentation of hazard analysis priorities based on a rationale which includes:

- Extent of process hazards
- Number of potentially affected employees
- Age of the process
- Operating history of the process."

Another priority setting method is to examine the OSHA and first aid log and injury/illness reports to see what kinds of accidents are most prevalent. Use frequency measurements of at-risk behavior types if you have the information.

See Figure 7-1 for the relationship between controls and consequences. Set priorities according to the most significant problems. Note that controls are adequate when:

1. Written instructions such as JSAs on how to perform tasks safely are in-place.
2. A system is developed to address conformance and non-conformance.
3. Preventive measures are used to address measured non-conformance.

Analyze the severity of the hazards associated with your operations to determine the degree of risk. If conformance is high, risk is low and consequences are unlikely (Figure 7-1). By making conformance observable and measurable, controls can be applied to reduce risk before any consequences occur.

Measure the priority activities by inspection, behavioral observation, or audit. Audits are more comprehensive than inspections and require enough data to be statistically significant.

Achieving Employee Agreement

Employee buy-in is best accomplished through participation while setting up standards, priorities, and inspections/procedures to improve safety. Communicate the vision during staff meetings, plant tours, monthly safety committee meetings, and safety action team meetings.

Establish Accountability

Accountability for safety should be established for all facility functions. Hold management accountable for providing a safe work environment and for implementing policies and procedures that foster safe work practices. On the hourly side, safety is a condition of employment. It is a huge mistake to put all of the responsibility on either management or hourly employees. Set priorities and objectives that are to be met for the month/year. State expected outcomes such as "We will mark all facility acid and caustic containers with hazard warnings using the NFPA 704 system by August 1." Measure the progress as percent-complete or "accomplished/not everything accomplished" by the end of the month.

Implement a Communication Plan

Prevention of injuries and illnesses depends on the active participation of all personnel. This takes feedback to improve the system and to encourage people to take an active role.

Key messages to emphasize consistently are:

*   We make products (or materials for use in products) that are valuable to our customers.
*   We will meet or exceed the requirements of all government regulations.
*   We will strive for continuous improvement in our safety, health, and environmental performance.
*   We want a safe and healthy work force.
*   We need to work safely to avoid the trauma of accidents and business interruption.

Be sure to make periodic reports of progress towards your goals. Also ask department heads to use a variety of media to report the status of their objectives.

Application of Measurements and Performance Indicators

Only after the accountability system has been designed, documented, and communicated, can we begin to measure conformance. Determine which programs to measure (Figure 7-1) by reviewing incident/accident investigation reports, past inspections and deficiencies, employee suggestions, audits, etc. Cost-justify the measurement system and make sure you have enough objective data to credibly evaluate your safety programs.

Example—New Measurement System

The author was finishing a presentation on accident and incident investigation when client managers were deciding what measurements to use. They chose lost time and recordable injuries right away, then added worker compensation claims and premiums. Suddenly, a manager asked, "Shouldn't we count first aid injuries too? Also, we have spent a lot of time and effort getting people to wear safety glasses. I think it's important to measure whether people are actually doing it. The bottom of the chart (Figure 7-1) says that use of PPE is one of the prevention activities we should be tracking." Without any further discussion, the group adopted his recommendations.

Example—Comparison of Two Companies

In addition to compiling accident-related results, measuring upstream (before the accident) accident control activities is especially important. (These upstream activities are sometimes called *performance indicators.*) Consider the following comparison between two companies.[10]

Company A has a poor safety record but has the same number of OSHA recordable injuries (6) as company B, which has an excellent safety record. Both companies have 100 employees. Company A has experienced one fatality. A total of 26 serious OSHA violations and 15 repeat violations have resulted in fines totaling $2.6 million.

A single safety coordinator is responsible for six production sites and is expected to cover human resource and environmental activities as well. Management is not involved in safety, which has no written policies. No accidents have been *reported* following a "commitment to safety" statement issued by the company.

Company B, also with 6 recordable injuries, has a continually improving safety record. A measurement system is in place and a number of safety initiatives are driving the OSHA recordable rate slowly downward. An involved management team has upstream measures in place that demonstrate the controls are working.

Although the number of OSHA recordables were the same, Company A had little or no upstream activities to positively impact their safety record. Company A's insurance rates are four times higher than those for B. Direct and indirect costs associated with accidents and incidents are higher.

Employees in Company B are consistently reminded of safety and participate in the behavioral safety program by making observations. The company also holds contractors accountable for safety, requiring them to follow the same safety policies and procedures as employees. Yet if OSHA recordables were the only measure, Company A and Company B would be the same. Thus, depending upon the measurement, results-oriented measurements can be very misleading when making comparisons. Furthermore, they do not tell the whole story in terms of total safety performance. Results are not manageable. A month or two without accidents may give the illusion of control. However, when the injury rate goes up again, there is no indication of what when wrong with the system.

OSHA requires that recordable (cases that require medical treatment) and lost-time injuries and illnesses be documented on the facility OSHA 300 log. Almost all OSHA inspections will involve a detailed audit of this injury and illness log. Many so-called *compliance directives* (OSHA's directions to compliance officers) are based on these traditional measurements. Results measures are here to stay but readers are advised not to rely on them to guide day-to-day accident prevention activities.

Performance Indicators for Chemical Safety Management

Choose measures and performance indicators that reflect your organizational values. For chemical safety management, consider upstream measures (before the incident) and downstream measures (after the incident) related to

- Process incidents and near-incidents
- Records of spills or loss of containments
- Process safety systems installed such as pressure relief, spill containment, and other detection/safety interlock controls

- Use of hazard analysis and control systems such as process hazards analysis (PHAs), pre-start safety reviews (PSRs), management of change (MOCs), etc.
- Employee queries and reviews of material safety data sheets (MSDSs)

A variety of leading and trailing performance indicators with emphasis on the upstream measures should be selected (Table 7-1).

Evaluating a Safety Management System

Good safety programs involve the entire organization in setting realistic goals for safety performance. Then individuals are held accountable for achieving the goals. Accountability has no meaning without measurement. Starting with accountability—the most important factor in achieving good line safety performance—it becomes clear that this process really entails ways to better measure the line manager (and "measurement has long been safety's single biggest problem."[11])

Historically, results such as numbers of accidents, injury frequency rates, and lost time injuries have been used to evaluate safety programs. Results measures remain in use today, even though they offer little help in actually changing performance. There are at least two ways that do a better job of evaluating the effectiveness of the safety management system.

Perception Surveys

Use of perception surveys, though somewhat subjective, is a good way to determine the effectiveness of the management system on the safety culture. Answers to the following questions can help improve employee ownership of the safety system:

- Has a survey with statistically valid outcomes been conducted within the last year?
- Have survey results been communicated to employees?
- Have employee interviews been conducted to define concerns identified in the survey?
- Have areas of concern been addressed and if so, which ones?
- Have employees been trained to conduct surveys and/or interviews related to safety?
- What do employees know about accidents, incidents, OSHA recordable injuries, chemical hazards, etc.?

Up-stream	Down-stream	Performance indicator	Up-stream	Down-stream	Performance indicator
	x	Process-related incidents and near-incidents		x	Chemical spills and loss of containment
x		Safety systems installed: pressure reliefs, automatic isolation, spill containment, etc.	x		Hazard control methods used- PHAs, PSRs, MOCs, MSDSs, etc.
x		Total WC costs	x		HAZCOM & PSM audits
x		Average WC cost per man hour	x		Employee participation
	x	OSHA 200 Logs	x		Employee housekeeping
	x	Industry ranking		x	Injury illness reports on time
x	x	Benchmarking other companies	x		Job Safety Analyses
x		Employee perception surveys	x		Written work procedures
		% safe behaviors	x		Air sampling program
	x	Injury/illness frequency rates	x		Toxic exposure monitoring audits
	x	Injury/illness severity rates	x		Housekeeping audits
x	x	Investigations completed on time	x		Air sampling program
x	x	Corrective action plans implemented	x		Emergency drills as planned
x		Safety meetings held as scheduled	x		Safety & health spending per employee
x		Agenda provided in advance	x		Employees trained in CPR/first aid
x		Safety records updated and posted		x	Accidental spills & releases
x		Inspections completed on time	x		Community outreach activities
x		Inspection items closed	x		Off-the-job safety initiatives
x		Management safety communications	x		Reports of peer support for safety
x		Management safety participation	x		Certifications of safety & health staff
x	x	Near-accident reports	x		Percent safety goals achieved
x		Self-inspections and audits	x		Training conducted as scheduled
x	x	Contractor safety record	x		Training test scores
x		Contractor safety activities	x	x	OSHA audits- no citations
x		Employee safety complaints		x	Dollar amount of penalties
x		Employee safety suggestions		x	Average time to abate hazard

Table 7-1 Example List of Performance Indicators

Although individual questions may be highly subjective, taken as a whole over an entire work force, perception surveys become a valuable quantitative tool to assess safety. Differences between employee and management perceptions are particularly important to identify and quantify in order to assess the effectiveness of a safety program.

Audit Systems

Audits are another basis for an effective measurement system if they are well correlated to accident records. Dan Petersen cites an audit study in which equal weight was given to four factors: management accountabilities, OSHA compliance, proper paperwork and physical inspections, and employee interviews.[12] The two parts that had high accident correlation were accountabilities and employee interviews. The other two factors had a high negative correlation, and thus cancelled the two effective measures. By reducing the weighting of OSHA compliance and physical conditions, the audit became more meaningful.

James Manzella provides five criteria to evaluate a measurement system:

"1. The measurement must provide adequate data to evaluate the system.

2. Data gathered must be objective.

3. The measurement system must be credible, with personnel providing, analyzing and reviewing data.

4. The system must be cost justified.

5. Only well-documented systems can be objectively measured."[13]

Conclusions

The main point of this chapter is to integrate management systems for safety, quality, and productivity, and to measure what is important. Anything worth doing is worth measuring. Setting measurable goals is the basis for decision-making and continuous improvement. There are two important links to make when establishing goals and policy: connecting written company/facility policy with regulatory standards and linking policy/procedure non-compliance to increased risk.

Too often, the safety coordinator is assigned or assumes responsibility beyond the scope of their duties. Clearly establish accountability to the line organization early in the process. Balance leadership expectations for supervisors and managers by encouraging all employees to participate in the safety improvement process.

Two of the most common mistakes made by facilities are:

1. Failure to document and maintain safety policy and procedures
2. Failure to write an annual safety and health plan

Facilities need to maintain current policies and procedures in a safety and health manual. An annual plan with specific goals that include identification and controls of hazards will accomplish as much improvement as any other component of the safety program. Although the target is zero accidents, goals should deal with hazard identification and control, not just numbers.

Gathering and improving process safety systems and information is an essential element of chemical safety management. This includes extensive use of hazard and safety information coming from material safety data sheets, job safety analysis, and process hazards analysis.

The benefits of a strong safety accountability and measurement system are numerous. Key among these benefits are:

*   Better process understanding
*   Less downtime, waste, and direct/indirect costs due to accidents
*   Better problem identification and decision making
*   Better accident prevention and identification of potential hazards
*   Better communication resulting in improved individual and team performance
*   Better standards and performance throughout the organization

References

1. Dan Petersen, "What Measures Should We Use, and Why? Measuring Safety System Effectiveness" *Professional Safety*, American Society of Safety Engineers, October 1998, Figure 1.
2. Modified from James C. Manzella, "Measuring Safety Performance To Achieve Long-Term Improvement," *Professional Safety*, American Society of Safety Engineers, September 1999. Material used with permission.
3. *Plant Guidelines for Technical Management of Chemical Process Safety*, CCPS of the American Institute of Chemical Engineers, 1992, p. 16.
4. *Plant Guidelines for Technical Management of Chemical Process Safety*, p. 17.

5. "An Employer's Guide to Developing A Workplace Accident and Injury Reduction (AWAIR) Program", Minnesota Department of Labor and Industry, Occupational Safety and Health Division, June, 1998.

6. *Plant Guidelines for Technical Management of Chemical Process Safety*, p. 18.

7. *Plant Guidelines for Technical Management of Chemical Process Safety*, p. 21.

8. Ethanol 2000, LLP, Safety Management System (SMS).

9. James C. Manzella, "Measuring Safety Performance to Achieve Long-Term Improvement," *Professional Safety*, American Society of Safety Engineers, Sept. 1999.

10. The following comparison is adapted from Daniel Patrick O'Brien, "A Quantitative Approach to Safety Performance," *Professional Safety*, American Society of Safety Engineers, Aug. 1998, p. 41.

11. Dan Peterson, "What Measures Should We Use, and Why? Measuring Safety System Effectiveness," *Professional Safety*, ASSE, Oct. 1998.

12. Peterson, p. 38.

13. Manzella, p. 35.

Regulatory Requirements

Introduction

This chapter continues to build a facility infrastructure for chemical safety management. Our purpose and scope is to answer four key questions:

1. How does the facility safety manager find applicable OSHA, EPA, and DOT regulations?
2. How should these minimum regulatory requirements shape company and facility standards?
3. What are the implications of converting external standards into internal standards?
4. What are important considerations when integrating external codes and standards into internal policies and procedures?

Unfortunately, too many organizations are only concerned with safety measures required by the law. To make matters worse, concern is often not raised until a facility is in trouble. Safety coordinators must help management understand that, in the long run, it is not helpful to use OSHA regulations as the basis for a safety and health program. Plant management wants a system that depends on internal performance standards that meet or exceed external regulatory requirements. Therefore, fit OSHA and other regulatory requirements into a well-managed chemical safety management system, not the other way around.

Use of external standards enables organizations to make an important transition in their drive for continuous improvement in safety performance. The ultimate goal, however, is to move from being externally driven to having an internal driving force based on company/facility policies and procedures.

Some benefits of integrating external codes and standards into facility policy are the following.[1] Using these standards

- Promotes uniform safe work practice industry-wide
- Allows sharing of a wider experience base
- Provides a means for development of consensus

- Gives a legally defensible criteria on which to base designs

This chapter will provide an overview of regulations applicable to chemical safety management:

- Title 29 Code of Federal Regulations (CFR)—OSHA
- Title 40 Code of Federal Regulations (CFR)—EPA
- Title 49 Code of Federal Regulations (CFR)—DOT

Introduction to OSHA Standards

OSHA relies on a three part strategy to reduce injuries and illnesses: enforcement, compliance assistance, and partnerships.[2] As OSHA administrator John Henshaw points out, "Just because there is a regulation [doesn't mean] everyone will comply."

However, OSHA regulations are an important source of potential policy and procedures for chemical safety management. In this section, we will concentrate on applicable OSHA standards and provide examples of how to apply them to developing company policy.

Health and Safety Managers should maintain current volumes of the Code of Federal Regulations (29 CFR 1910 Labor) on their bookshelf and/or in the facility safety library. Current information about the Occupational Safety and Health Administration is best found on OSHA's home page, www.osha.gov. These standards do not change frequently, but it is a good idea to stay current on interpretation of the code, examples, and resources.

Safety and health coordinators should be extremely careful about investing time in non-government sources that claim to summarize or simplify safety and health codes. Some of these sources provide excellent systems and tools for implementing OSHA requirements, but there is no substitute for starting with company goals and listing all current "musts and shalls" in the title 29 Code of Federal Regulations (CFR) part 1910 Occupational Safety and Health Standards.

OSHA Consultation Service

The OSHA consultation service is free to employers, funded by OSHA, and delivered by many state OSHA offices. Check the OSHA home page to determine whether your location has a state plan. This service is completely separate from inspection/enforcement and is primarily intended for small business organizations that often do not have professional safety and health services readily available. Thus OSHA consultation is very helpful to the small chemical processing facility that doesn't have a strong safety and health infrastructure.

OSHA's small business program provides excellent resources for the smaller plant. Data and other services include locating OSHA inspections by name of company, frequently cited standards, profiles of citations by industry, recordkeeping guidelines for small business, and other inspection and statistical information.

OSHA Inspection and Citation Data

Studying OSHA violation data is useful for setting a priority for company standards. The OSHA web page includes industry profiles of inspection, citation, and penalty data. Typing in OSHA standards that apply

Table 8-1 Inspection and Citation Data for Chemicals and Allied Products, 1999-2000[3]

1910 Standard	# Cited	#Inspections	$ Penalties	Rank
Process Safety Management				
1910.119 All industry	588	102	$3,701,282	
Division D Mfg.	509	79	3,468,210	
2800 Chemical and Allied	233	32	2,876,868	1
Average $ per citation			$12,347	
Hazardous Waste Operations and Emergency Response				
1910.120 All industry	399	176	$381,188	
Division D Mfg.	255	113	$274,404	
2800 Chemical and Allied	82	31	$153,706	1
Average $ per citation			$2077	
Flammable and Combustible Liquids				
1910.106 All industry	1015	615	$670,628	
Division D Mfg.	779	453	573,786	
2800 Chemical and Allied	74	35	109,726	2
Average $ per citation			$1482	
Confined Space Entry				
1910.146 All industry	1534	539	$1,568,516	
Division D Mfg.	1051	398	973,301	
2800 Chemical and Allied	85	23	104,201	6
Average $ per citation			$1225	
Lockout Tagout				
1910.147 All industry	4149	2231	$3,741,255	
Division D Mfg.	3279	1732	3,217,970	
2800 Chemical and Allied	129	66	22,603	10
Average $ per citation			$175	
PPE Eye-Face				
1910.133 All industry	588	565	$431,391	
Division D Mfg.	380	361	308,744	
2800 Chemical and Allied	17	17	14,844	12
Average $ per citation			$873	
Hazard Communication				
1910.1200 All industry	7421	3802	$1,537,911	
Division D Mfg.	3334	1721	884,085	
2800 Chemical and Allied	177	85	124,023	14
Average $ per citation			$700	

to chemical safety will generate a chemical industry profile of penalties (Table 8-1). The data is presented in order of how the chemical and allied product industry ranks with other industries in its SIC group.

The chemical industry (standard industrial classification 2800) ranks number one among other groups for process safety management citations. This may be due to the standard's broad scope and to the effort of special inspectors who sometimes spend weeks auditing chemical facilities. Other industries in 1999-2000 with high PSM rankings include:

2)   Food and Kindred Products
3)   Petroleum Refining and Related Industries
4)   Motor Freight Transportation and Warehousing
5)   Primary Metal Industries

What may be surprising to the reader is the relative size of penalties for PSM violations compared to those for the other standards. The average penalty ($12,347) for PSM is 17 times greater than that for HAZCOM ($700). This is because HAZCOM violations are often other-than-serious (maximum penalty per violation of $7,000) and generally do not affect many plant employees. However, many PSM violations are serious (same maximum penalty per violation) with potentially catastrophic consequences— so that the number of employees in the plant may multiply the penalty!

Other-than-serious violations are those that have a direct relationship to job safety and health, but probably would not cause death or serious physical harm. Serious violations are those where there is substantial probability that death or serious physical harm could result and that the employer knew, or should have known, the hazard existed. A willful violation (penalties of up to $70,000 each) is one that the employer intentionally and knowingly commits or a violation that the employer commits with indifference to the law. Here, the employer either knows that a violation has occurred, or knew that the hazardous condition existed and made no reasonable effort to eliminate it.

Table 8-2 lists process safety violations by paragraph for the five-year period 1992-1997.

The top four violations (in order by number of violations) are operating procedures, mechanical integrity, process hazards analysis, and process safety information. This result is not surprising, given that so many plants operate with no or inadequate written operating procedures and do not give enough attention to inspection, test, and maintenance activities.

Table 8-2 Process Safety Violations by Paragraph[4]

1992-1997 Standard		Total Violations	Rank
1910.119C-	Employee Participation	442	5
1910.119D-	Process Safety Information	569	4
1910.119E-	Process Hazards Analysis	658	3
1910.119F-	Operating Procedures	925	1
1910.119G-	Training	429	6
1910.119H-	Contractors	366	8
1910.119I-	Pre-start Safety Review	56	12
1910.119J-	Mechanical Integrity	758	2
1910.119K-	Hot Work Permit	61	11
1910.119L-	Management of Change	374	7
1910.119M-	Incident Investigation	159	9
1910.119N-	Emergency Planning and Response	140	10
1910.119P-	Trade Secrets	1	13
Total		4938	

General Duty Clause

Each subpart of the 1910 standard contains pages and pages of detailed OSHA code (Table 8-3). In spite of this scope and detail, there are still many plant hazards not covered by a specific standard. OSHA's general duty clause is used to cover violations where no specific standard exists.

The requirements of specific standards always take precedence over the general duty clause; however, the general duty clause can be used in situations where an OSHA inspector determines that the intent of the law is being violated, but does not have specific language to back it up.

> Each employer.... shall furnish to each of his employees employment and a place of employment which is free from recognized hazards that are causing or are likely to cause death or serious physical harm to his employees.

Figure 8-1 OSHA General Duty Clause[5]

Example

An OSHA inspector comes to a chemical plant due to an employee complaint about an acid transfer system. A near accident occurred during a gravity transfer operation between a full tote tank outside the building into an empty tank inside the building. A faulty coupling broke during the transfer, resulting in a stream of acid rushing by the operator's face, splashing against the wall, and flowing across the floor. The inspector takes photos of the faulty equipment and notes that the written procedure calls for use of a pump to transfer the acid. There is no specific standard covering the equipment and procedure. Thus, the inspector cites the employer under the general duty clause.

Note that the General Duty Clause is Section 5(a) (1) of the actual OSH Act of 1970. Because it was authorized by Congress to cite employers right away for violations, it is *not* found in the Code of Federal Register along with other regulating standards.

OSHA Standards Applicable to Chemical Safety Management

Table 8-3 OSHA Standards Related to Chemical Safety and Health

1910 Sub.	Subject	Relevance to Chemical Safety and Health
G	Occupational Health and Environmental Controls	Covers ventilation requirements and exposure to noise, and non-ionizing radiation.
H	Hazardous Materials	Covers general requirements for compressed gases and specifics for acetylene, hydrogen, oxygen, nitrous oxide and flammable/combustible liquids. Defines safe work conditions and procedures for hazardous spray finishing, dip tanks, explosives and blasting agents, storage and handling of liquefied petroleum gases and anhydrous ammonia. Also covers ProcessSafety Management of Highly Hazardous Chemicals (PSM) and Hazardous waste operations and emergency response (HAZWOPER).
I	Personal Protective Equipment (PPE)	Provides general requirements (includes need for hazard assessment) and specific needs for eye and face, respiratory, head, foot, and hand protection. Also defines requirements for electrical protective equipment.

Table 8-3 OSHA Standards Related to Chemical Safety and Health (cont.)

1910 Sub.	Subject	Relevance to Chemical Safety and Health
J	General Environmental Control	Includes sanitation, safety color code for marking physical hazards, specifications for accident prevention signs and tags, permit-required confined spaces, and control of hazardous energy (lockout/tagout requirements include chemical line-breaking).
K	Medical and First Aid	Policies and procedures may be developed by reviewing specific Material Safety Data Sheets (MSDSs) for chemicals stored and handled.
L	Fire Protection	Covers fire brigades, fixed fire suppression equipment and other fire protection systems. Helps facilities to develop responsibilities and duties and to define equipment needed for an emergency squad.
N	Materials Handling and Storage	Calls out requirements for powered industrial trucks in hazardous locations (near compressed gases, flammable and combustible liquids)
Q	Welding, Cutting, and Brazing	Defines specific requirements for hot work near flammable and combustible liquids. Also see 1910.119 (k)- Hot work permit.
R	Special Industries	Includes pulp, paper and paperboard mills (some covered by PSM); textile industry; bakery equipment; laundry machinery and operations; sawmills and logging operations; agricultural operations; telecommunications; electric power generation, transmission and distribution; and grain handling facilities (explosive dust).
S	Electrical	Covers design safety standards for electrical systems. This includes classified (flammable materials) locations. Also covers safety-related work practices, maintenance requirements and requirements for special equipment.
Z	Toxic and Hazardous Substances	Covers air contaminants in general and many specific chemicals, blood borne pathogens and hazard communications (HAZCOM) or employee-right-to-know (RTK) requirements. HAZCOM/RTK is an important link to PSM- process safety information). Also includes retention of DOT markings, placards and labels, and chemical exposure in laboratories.

Table 8-3 lists the 1910 subparts that relate to chemical safety and health. Note that there is considerable overlap between the requirements of many subparts. For example, see the requirements for welding, cutting, and brazing near flammable and combustible materials (subpart Q). Subpart H, 1910.119 (Process Safety Management) requires a permit to authorize what OSHA calls "hot work," covering many of the same points in subpart Q.

Subparts Z and H form another important overlap. Much of the information contained in material safety data sheets (subpart Z) is required in the Process Safety Management standard as Process Safety Information.

The three OSHA standards most applicable to chemical safety management are:

- Hazard Communication (HAZCOM)
- Process Safety Management (PSM)
- Hazardous Waste Operations and Emergency Response (HAZWOPER)

Hazard Communication

Figure 8-2 is an outline of the1910.1200 Hazard Communication or HAZCOM Standard.

---

a) Purpose
b) Scope and application
c) Definitions
d) Hazard determination
e) Written hazard communication program
f) Labels and other forms of warning
g) Material safety data sheets
h) Employee information and training
i) Trade secrets
j) Effective dates

Memory-aid Shortcut:
T   stands for **training**
I   stands for chemical **inventory**
P   stands for written **program**
S   stands for Material Safety Data **Sheets**

Figure 8-2 OSHA Hazard Communication Standard

Item (g) (material safety data sheets) is sometimes called the "heart" of the HAZCOM Standard because MSDSs provide facilities with the following important safety and health information about the chemicals with which they work:

- Product information—chemical and product name, manufacturer, emergency phone numbers, permissable exposure limits (PELs), as well as potential physical and health hazards.

- Exposure situations—first-aid measures, spill or leak procedures, and fire-fighting procedures.

- Hazard prevention and protection—how to handle and store materials to prevent or minimize direct contact, fire hazards, release or loss-of-containment, and hazardous conditions such as excessive heat, direct sunlight, or vibration when handling the material. Other topics covered are engineering controls, PPE required, physical and chemical properties, stability of the material, conditions and materials that cause dangerous reactions with the chemical, and hazardous decomposition products.

- Other specific information—may include toxicological, ecological, disposal, transportation, and regulatory information needed by emergency responders, doctors, and other emergency specialists.

Facility safety coordinators must be alert for opportunities to use OSHA and other standards to improve safe work practices on the factory floor. Use of PPE is a good example. Appropriate PPE for a given chemical is always specified in the MSDS, but that is no guarantee for compliance. However, by implementing a hazard assessment (CFR 1910.132(d)), an OSHA PPE requirement, you can get the practice to match your policy. Figure 8-3 is an example of a hazard assessment checklist. Subpart I, Appendix B provides non-mandatory compliance guidelines for hazard assessments and personal protective equipment selection.

Complete the checklist for each department with the assistance of an area supervisor or lead operator. Then help plant management draft a PPE and discipline policy statement to reinforce use of personal protective equipment.

Process Safety Management Standard

After a series of process-related catastrophes in the U.S. and abroad, OSHA proposed the process safety management standard (PSM), 29 CFR 1910.119. The final rule was made effective May 26, 1992, and all ele-

Department: Maintenance   Date: August 28, 2001, modified December 20, 2001

Procedure	Hazard Category	PPE #							Notes
		1	2	3	4	5	6	7	
Lockout-Tagout	All	X	X	X	X	X	X	X	Goggles with face shield
Line Breaking	All	X	X	X	X	X	X	X	Goggles with face shield
Welding	E,F,H,I,L,N	X		X	X		X	X	Suitable filter lenses
Cutting, grinding	E,F,H,I,L,N	X		X	X		X	X	Suitable filter lenses
Electrical	E,F,H,I,L	X		X	X	X	X		Electrical insulated gloves, shoes
Mechanical	All	X	X	X	X	X	X	X	Evaluate hazards
Confined space entry	All	X	X	X	X	X	X	X	Initial and periodic air monitoring

HAZARD CATEGORIES

C   Compression (e.g. equipment roll-over)
Ch   Chemical (see MSDS)
D   Dust
E   Electrical shock
F   Flammable
H   Heat
I   Impact
L   Light (optical) radiation
N   Noise
P   Penetration

**PPE CATEGORIES
(evaluate specific hazard)**

1. Eye and face protection
2. Respiratory protection
3. Head protection
4. Foot protection
5. Electrical protective equipment
6. Hand protection
7. Hearing protection

Assessment by: _____   Signature: _____

Figure 8-3 Hazard Assessment for Personal Protective Equipment (PPE)[6]

ments of the standard were required to be in place and functioning by May 26, 1997. Government and industry representatives designed the comprehensive standard to guide facilities in the safe management of processes that use "highly hazardous chemicals." The main purpose of PSM is the application of management systems to identify, understand, and control process hazards that will prevent on and offsite, process-related catastrophic events.

These events include large-scale releases of a variety of materials that are toxic, flammable, explosive or reactive, or that have a combination of harmful properties.

There are 137 toxic and reactive chemicals, certain flammables, explosives, and pyrotechnics on the list. Figure 8-4 is an outline of the standard found in subpart H of the OSHA 1910 regulations.

(a) Application

(b) Definitions

(c) Employee participation

(d) Process safety information

(e) Process hazards analysis

(f) Operating procedures

(g) Training

(h) Contractors

(i) Pre-startup safety review

(j) Mechanical integrity

(k) Hot work permit

(l) Management of change

(m) Incident investigation

(n) Emergency planning and response

(o) Compliance Audits

(p) Trade secrets

APPENDIX A- LIST OF HIGHLY HAZARDOUS CHEMICALS, TOXICS AND REACTIVES

Figure 8-4 Process Safety Management Standard

Note the contrast between the HAZCOM standard (personnel safety, Figure 8-2) that specifies detailed methods of chemical hazard control and PSM, a performance standard that describes what must be done without defining in detail how it is to be done.

The PSM standard sets broad performance requirements for management systems that:

1. Identify process-related hazards
2. Estimate the risk of catastrophic effects
3. Assess the effectiveness of existing or as-designed controls

If existing safeguards are judged adequate, no additional action is taken. If not, additional measures are considered, including equipment, hardware, inherently safe systems (discussed in Chapter 3), and administrative controls.

Five PSM elements that work together to form a driving force for chemical safety management, called the engine of process safety management, include the following:

1. Process safety information
2. Process hazards analysis
3. Operating procedures
4. Pre-start safety reviews
5. Management of change

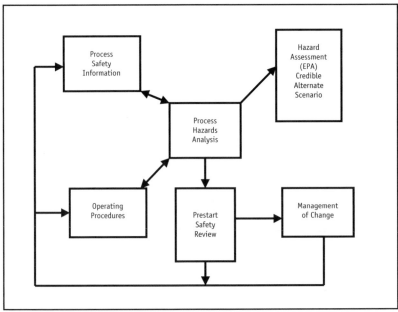

Figure 8-5 Engine of Process Safety Management

Safety coordinators must work with facility employees to develop the PSM engine incrementally. (PHAs were discussed in depth in Chapter 4 and Chapter 5. In Chapter 4 (see Figure 4-8), the PHA becomes an investigation tool for actual and potential incidents. Chapter 5 contains the actual "how to" for conducting PHA studies and compares the various techniques (see Table 5-3). The following sections explain the necessary information, analysis, and procedures to effectively integrate the PSM regulation into facility policy and procedures.

Process Safety Information

Process Safety Information (PSI) includes chemical, process, and equipment safety information. Note that well-documented PSI is critical to a safe process design. The OSHA standard requires that as much of the PSI is brought together before conducting the Process Hazards Analysis (PHA). For the list of OSHA PSI requirements, see Chapter 3.

Updated Process Safety Information is documented following the PHA, and both revision of operating procedures and the pre-start safety review should follow the documentation. Be sure to establish an effective management of change system early (Figures 5-6 and 5-7).

Think of PSI as a "living library" of information that is maintained for the life of the process. PSI provides significant operability benefits in addition to safety. The process understanding gained in compiling the information will help you considerably in reducing process variability.

Process Hazards Analysis

The PHA is a central tool for corrective action. The results of the PHA study drive changes in facilities, equipment, and procedures that reduce the risk of a catastrophic event.

OSHA requires that "The employer shall perform an initial process hazard analysis (PHA) or hazard evaluation on processes covered by this standard. The process hazard analysis must be appropriate to the complexity of the process (Table 5-3) and shall identify, evaluate, and control the hazards involved in the process." The hazards involved (even those involving human error) are caused by deviations from normal process conditions and procedures.

PHA studies take a lot of time and effort, and OSHA required facilities to establish a priority order for conducting PHAs based on consideration of things such as the extent of process hazards, number of potentially affected employees, as well as age and operating history of the process. Other PHA requirements are

- Employers shall use one or more of the methodologies that are appropriate to determine the hazards of the process: What-If, Checklist, What-If/Checklist, Hazard and Operability Study (HAZOP), Failure Mode and Effects Analysis (FMEA), Fault Tree Analysis, or an appropriate equivalent method.
- PHA studies must address:
  - ❏ hazards of the process
  - ❏ identification of previous incidents that had a likelihood of catastrophic consequences
  - ❏ engineering and administrative controls and their interrelationships
  - ❏ consequences of failure of engineering and administrative controls
  - ❏ facility siting
  - ❏ human factors
  - ❏ qualitative evaluation of a range of the possible safety and health effects of failure of controls on employees
- Performance of the PHA by a team with expertise in engineering and process operations, including one member knowledgeable in the specific PHA methodology being used.
- Use of a system to promptly address, resolve, and document the team's findings and recommendations.
- Update and revalidate the studies at least every five years.
- Retain PHA studies, updates, and revalidations for the life of the process. Maintain this documentation as well-organized process safety information. Do not place memos, copies of revised operating procedures, etc. into a file with the original PHA.

Operating Procedures

Many covered PSM facilities fall short of the OSHA requirements for well-written operating procedures. The procedures for safely conducting work need to address these elements:

- Steps for each operating phase, including:
  - ❏ initial startup
  - ❏ normal and temporary operations
  - ❏ emergency shutdown procedures

- ❏ emergency operations
- ❏ normal shutdown and startup following a turnaround or emergency shutdown
- Operating limits with consequences of deviation and steps required to correct or avoid the deviation
- Safety and health considerations involving properties and hazards of chemicals used in the process, including:
  - ❏ exposure prevention measures (including engineering controls, administrative controls, and personal protective equipment)
  - ❏ control measures to be taken if exposure occurs
  - ❏ quality control for raw materials
  - ❏ control of hazardous chemical inventory levels
  - ❏ measures and/or controls for any special or unique hazards
- Safety systems and their functions, including:
  - ❏ operating procedures readily accessible to employees who work in or maintain the process
  - ❏ a system for updating operating procedures to ensure that they reflect current operating practice, including modifications that result from changes in process chemicals, technology, equipment, or facilities
  - ❏ annual certification to ensure operating procedures are current and accurate
- Safe work practices, including:
  - ❏ lockout-tagout
  - ❏ confined space entry
  - ❏ opening process equipment or piping (line breaking)
  - ❏ control over entrance into the facility by maintenance, contractor, laboratory, or other support personnel

Many facilities have documented procedures for normal process operation but need to work on the other operating phases required above. Another significant challenge to process safety coordinators is to assure that what is described in the procedure is actually done on the factory floor.

Pre-start Safety Review

Consider the pre-start review the final check before the operating crew starts pushing buttons and opening valves.

OSHA's minimum requirements for pre-start reviews (PSRs) are that facilities have:

- Performed a PSR for new facilities and for modified facilities where the modification requires changes to the process safety information;
- Confirmed that construction and equipment is in accordance with design specifications;
- Verified that safety, operating, maintenance, and emergency procedures are in place and are adequate;
- Confirmed that a PHA has been performed and recommendations have been resolved or implemented before startup of new facilities;
- Confirmed that modified facilities meet the management of change requirements; and
- Ensured that *training* for employees involved in the process has been completed.

In addition, use the pre-start review for any non-covered processes where safety or operability is a concern. One facility required laboratory and process development personnel to complete a pre-start review before using the plant's equipment for any factory experiment. This was done regardless of the material's regulatory status.

Management of Change

OSHA's management of change requirements are stated simply but considerable effort is required to establish an effective system. Implement:

- Written procedures to manage changes (except replacements in kind) to process chemicals, technology, equipment, and procedures as well as changes to facilities that affect a covered process. Include:
  - ❑ Technical basis for the proposed change
  - ❑ Impact of change on safety and health
  - ❑ Modifications to operating procedures
  - ❑ Necessary time period for change
  - ❑ Authorization requirements for the proposed change
- Training for affected employees including contractors prior to startup.
- A system for documenting new process safety information.
- A system for updating operating procedures or practices as necessary.

Appendix C for 1910.119 has guidelines for complying with PSM. This performance standard is operational, meaning that process and manufacturing personnel must implement it. Gain an understanding of the basic requirements before contacting an external or internal consultant to help direct the work.

Hazardous Waste Operations and Emergency Response Standard

The Hazardous Waste Operations and Emergency Response (HAZWOPER) Standard applies to a broad range of industries in which hazardous substances are used. It also regulates how emergency response plans are drafted and the requirements for hazardous materials (HAZMAT) team activities at industrial sites.

These so-called hazardous substances include common chemicals, such as gasoline. The requirements are found under Subpart H, 1910.120 (Figure 8-6).

(a) Scope, application and definitions

(b) Safety and health program

(c) Site characterization and analysis

(d) Site control

(e) Training

(f) Medical surveillance

(g) Engineering controls

(h) Monitoring

(i) Informational programs

(j) Handling drums and containers

(k) Decontamination

(l) Emergency response by employees at uncontrolled hazardous waste sites

(m) Illumination

(n) Sanitation at temporary workplaces

(o) New technology programs

(p) Certain Operations Conducted Under the Resource Conservation and Recovery Act of 1976 (RCRA)

(q) Emergency response to hazardous substance releases

APPENDIX A- PERSONAL PROTECTIVE EQUIPMENT TEST METHODS

APPENDIX B- GENERAL DESCRIPTION AND DISCUSSION OF THE LEVELS OF PROTECTION AND PROTECTIVE GEAR

APPENDIX C- COMPLIANCE GUIDELINES

APPENDIX D- REFERENCES

APPENDIX E TRAINING CIRRICULUM GUIDELINES

Figure 8-6 HAZWOPER Standard

Covered Facilities and Substances

Chemical plants and refineries are the main target facilities for the HAZWOPER Standard. However, coverage is much broader than might be expected. The standard applies to five groups of employers and their employees. This includes any employees who are exposed or potentially exposed to hazardous substances—including hazardous waste—and who are engaged in one of the following:[7]

- Clean-up operations—required by a governmental body and conducted at uncontrolled hazardous waste sites;
- Corrective actions involving clean-up operations at federal sites covered by the Solid Waste Disposal Act and amended by the Resource Conservation and Recovery Act of 1976 (RCRA);
- Voluntary clean-up operations at sites recognized by a governmental body as uncontrolled hazardous waste sites;
- Operations involving hazardous wastes that are conducted at treatment, storage, and disposal facilities regulated by EPA or by agencies under agreement with EPA to implement RCRA regulations; and
- Emergency response operations for releases of, or substantial threats of release of, hazardous substances regardless of the location of the hazard (includes industrial facilities).

Beyond chemical plants and refineries, few industries escape these requirements. For example, food and pharmaceutical industries use ethylene oxide to sterilize their products and thus have hazardous waste. They may also use chlorine to disinfect surfaces and anhydrous ammonia to refrigerate intermediates and products.

The HAZWOPER code calls out harmful biological agents and hazardous substances covered by OSHA (Subpart Z—list of toxic and hazardous substances), EPA (covered by Superfund and other regulations), and DOT (materials from HAZMAT tables).

Incidental versus Emergency Release

An incidental release is one which does not pose a significant safety or health hazard to employees in the immediate vicinity or to the employee cleaning it up, nor does it have the potential to become an emergency within a short time frame. Incidental releases are limited in quantity, exposure potential, or toxicity and present minor safety or health hazards to employees in the immediate work area or those assigned to clean them up. An incidental spill may be safely cleaned up by employees who are familiar with the hazards of the chemicals with which they are working.

An emergency release is an uncontrolled release that is likely to require an outside response. This includes responders such as mutual-aid groups, local fire departments, and off-site emergency squads. Distinction between incidental and emergency is determined by the facility, and specific procedures for both are included in the emergency response plan.

HAZWOPER Compliance Strategies

There are basically three ways to comply with the HAZWOPER standard based on knowing the types and amounts of the hazardous materials:

1. Evacuation—this strategy requires the facility to develop an evacuation plan, identify critical shutdown procedures, provide evacuation plan training (basically a periodic drill providing HAZCOM and first responder awareness level training), and coordinate with a local emergency response group.
2. Contain or confine the spill—this is a defensive choice that does not require a HAZMAT team. Required elements are an emergency response plan, training at the first responder operations level, provisions for protective clothing, establishing emergency response procedures, and regular response training.
3. Aggressively abate the spill—this strategy requires a HAZMAT team and an emergency response plan and incident command system, training at the HAZMAT technician level, provisions for medical surveillance of emergency responders, chemical protective clothing and procedures, as well as regular training.

Facility safety and health coordinators should concentrate on two key parts of the standard before drafting a plan or making compliance recommendations to plant management:. The first is the written program requirements. The second key part, 1910.120 Appendix C—Compliance Guidelines, provides guidance on how to comply effectively and efficiently with the regulations. The Compliance Guidelines are organized into nine sections.

Occupational Safety and Health Program

Before HAZWOPER standards can be implemented, a functioning safety and health program and a site coordinator is necessary. Note that a written hazard communication program (HAZCOM), discussed above, is the cornerstone for the overall safety and health program.

The OSHA guidelines identify six key features for the program:

1. A policy statement stating lines of authority and accountability for implementing the program;
2. A system for developing hazard identification and control procedures, including accident investigation;
3. A communication plan and a system for employee involvement;
4. Training for supervisors and employees;
5. Provisions for emergency planning and response; and
6. Methods to evaluate and continuously improve the program through information feedback.

The guidelines underscore the need for professional safety and health staff to develop and implement the site plan. See Chapter 10 for more information on qualified safety, health, and environmental staffing.

Training

The training guidelines for those involved in HAZWOPER incidents describe a combination of classroom and hands-on instruction. Annual training content is described generally for employees, HAZMAT specialists, incident commanders, technical experts, medical experts, and support people.

Those who respond to HAZMAT incidents need to be trained at one of five levels:[8]

Level 1: First Responders at the Awareness Level
Level 2: First Responders at the Operations Level
Level 3: Hazardous Materials Technician
Level 4: Hazardous Materials Specialist
Level 5: On-Scene Incident Commanders

Note that the OSHA guidelines do not clearly identify these five levels. Seek materials from a HAZWOPER training provider or contact a competent consultant before making a decision on how to proceed with training.

In addition, OSHA suggests two NFPA consensus standards for study: NFPA 471—"Recommended Practice for Responding to Hazardous Material Incidents" and NFPA 472—"Standard for Professional Competence of Responders to Hazardous Material Incidents."

Most facilities, fire departments, and other emergency response organizations use subject matter experts to deliver the training. A typical agenda involves one day for classroom training and one day for an exercise using full PPE.

Decontamination

A major safety consideration is to establish a method to assure that employees are not exposed to chemicals by re-using contaminated PPE. OSHA provides detailed references to EPA and Coast Guard decontamination information.

Emergency Response Plans

The OSHA guidelines stress that employers need to be sure the site plan is compatible with state, district, and local emergency response plans. NRT-1, the *Hazardous Materials Emergency Planning Guide*, is the major reference. CMA's CHEMTREC and the *Fire Service Emergency Management Handbook* are also good sources of information. DOT's *Emergency Response Guidebook* is also an excellent reference for first responders during the initial phase of a HAZMAT incident.

PPE Programs

According to the guidelines, PPE must be suitable for protection against chemical, physical, and biological hazards. HAZWOPER coordinators should be aware of worker hazards created by the PPE itself. These include heat and other physical stress; psychological stress; and impaired vision, mobility, and communication. The written program must include hazard identification, medical monitoring, environmental surveillance, selection, use, maintenance, and decontamination of PPE. In addition, potential responders must be trained in these topics.

Incident Command System

The ICS system recommended by OSHA is similar to the fire service's popular command post setup. The idea here is that strategy, tactics, and all decision-making is done by *one* individual.

The OSHA guidelines provide two examples. First, procedures are described for a small incident such as an overturned tank truck with a small leak of flammable liquid. In this case, the first responding senior officer would determine all appropriate actions.

In the second example, involving a large and complex incident, some delegation is required in order to achieve a manageable span of control. Considerations that may be delegated include provision of medical services, evacuation, water supply, equipment resources, media relations, and safety and site control. Large incidents often require integration of activities with police for crowd and traffic control. Providing an organized system for back-up personnel is also critical for large incidents. It is staggering to think of the complex ICS that was used for the New York World Trade Center incident September 11, 2001.

Site Safety and Control Plans

Site safety and control plans must be part of the employer's emergency response plan. The guidelines basically stress the hazard and risk analysis required to protect all responders. The incident commander should have a site map or sketch at hand, indicating clean, transition, and decontamination zones. Use the buddy system to keep track of the workers.

Medical Surveillance Programs

A medical surveillance program is important as part of a HAZWOPER plan to assess and monitor the health of all workers. The guidelines remind coordinators to keep accurate records for future reference. OSHA provides sources of more detailed information on medical surveillance from the National Institute for Occupational Safety and Health (NIOSH), OSHA, and the U.S. Coast Guard.

New Technology and Spill Containment Programs

This section involves methods such as conventional diking and ditching to contain and control spilled chemicals. Users are encouraged to employ new materials and equipment that solidify liquids and suppress flammable vapors.

Once these nine sections have been consulted, communicate the facility HAZWOPER plan and describe how it is integrated with the overall safety and health plan. This includes naming the leader and members of the incident emergency response crew. Keep the plan on disk or on your server so that it is current. Also, draft site-specific standard operating procedures (SOPs) for HAZWOPER incidents. Incorporate emergency procedures from the site safety and health plan if you have them.

Consensus and Industry Standards

In 1970, OSHA was required by Congress to adopt many national consensus standards without change. Consensus standards (sometimes called teaching standards) come from groups of experts in their respective fields that come together for the purpose of generating detailed requirements for complex systems. These technical experts represent companies and other stakeholder organizations representing a particular industry. The Chemical Process Safety Report provides a comprehensive list of consensus organizations that develop standards relevant to the chemical process industry.[9] Six of the key organizations are:

- American National Standards Institute (ANSI): Experts in materials selection and design of piping, valves, flanges and fittings, and other process equipment;

- American Petroleum Institute (API): Generate recommended safety and process hazards management practices that govern the design of petroleum and other hydrocarbon processing facilities.

- American Society of Mechanical Engineers (ASME): Define requirements for the national boiler and pressure vessel code; welding materials and welder qualifications; specifications for ferrous and nonferrous materials.

- American Society for Testing and Materials (ASTM): Expertise in standard testing methods and acceptable results, as well as description of metallic and non-metallic materials.

- Institute of Electronics and Electrical Engineers (IEEE) and Instrument Society of America (ISA): Expertise in the design and application of electrical and electronic controls, instruments, and other equipment as well as failure rates for devices.

- National Fire Protection Association (NFPA): Expertise in electrical area classifications and requirements for fire protection systems.

Figure 8-7 shows additional external consensus organizations and applicable standards for chemical process safety.[10]

Note that OSHA can reference a consensus standard in a citation where the CFR code does not specify the requirement in detail.

NFPA 30, Flammable and Combustible Liquids Code, is a good example of a consensus standard often used by insurance inspectors and others to conduct what are called facility *Loss Prevention Surveys* (see Chapter 5).

American National Standards Institute (ANSI)

11 W. 42ⁿᵈ St., New York, N.Y. 10036

www.ansi.org

- ANSI K61.1-1989, Safety Requirements for the Storage and Handling of Anhydrous Ammonia.
- ANSI Y32.11-1961 (reaffirmed 1993), Graphic Symbols for Process Flow Diagrams for the Petroleum and Chemicals Industry.
- ANSI Z129.1-1994, Precautionary Labeling of Hazardous Industrial Chemicals
- ANSI Z400.1-1993, Hazardous Industrial Chemicals- Materials Safety Data Sheets- Preparation.

American Petroleum Institute (API)

1220 L St. N.W., Suite 900, Washington, D.C. 20005

www.api.org

Standards

- API 510-1992, Pressure Vessel Inspection Code-Maintenance, Inspection, Rating, Repair and Alteration, Order No. C510000.
- Std 610-1995, Centrifugal Pumps for Petroleum, Heavy Duty Chemical, and Gas Industry Services, Order No. C61008.
- Std 617-1995, Centrifugal Compressors for Petroleum, Chemical, and Gas Industries, Order No. C61706.

Recommended Practices

- RP 521-1990, Guide for Pressure-Relieving and De-pressuring Systems, Order No. C52100.
- RP 683-1993, Quality Improvement Manual for Mechanical Equipment in Petroleum, Chemical and Gas Industries, Order No. C68300.

American Society of Mechanical Engineers (ASME)

345 E. 47ᵗʰ St., New York, N.Y. 10017

www.asme.org

- 1992, Boiler and Pressure Vessel Code
- B 16.5-1988, Pipe Flanges and Flanged Fittings, Order No. MX1588.
- B 31.3-1993, Chemical Plant and Petroleum Refinery Piping, Order No. A 03793.
- B 31.4-1992, Liquid Transportation Systems for Hydrocarbons, Liquid Petroleum Gas, Anhydrous Ammonia and Alcohol, Order No. A03892.

American Society for Testing and Materials (ASTM)

100 Barr Harbor Drive

West Conshohocken, Pa. 19428

www.astm.org

Figure 8-7 Consensus Organizations and Standards

The following ASTM standards are available from ANSI:

- ANSI/ASTM D 581-1989, Practice FOR Determining Chemical Resistance OF Thermosetting Resigns Used IN Glass Fiber-Reinforced Structures Intended for Liquid Service (07.02).

- ANSI/ASTM D5592-1994, Guide for Material Properties Needed in Engineering Design Using Plastics (08.03).

The Chlorine Institute, Inc.

2001 L Street, N.W., Washington, D.C. 20036

www.cl2.com

- Pamphlet 5-1992, Non-Refrigerated Liquid Chlorine Storage.

- Pamphlet 86-1994, Recommendations to Chlor-Alkali Manufacturing Facilities for the Prevention of Chlorine Releases.

Institute of Electronics and Electrical Engineers (IEEE)

345 East 47th St.

New York, N.Y. 10017

www.ieee.org

- IEEE C2 88-90, National Electrical Safety Code Interpretations, 1988-1990 Inclusive.

Instrument Society of America (ISA)

67 Alexandra Drive, P.O. Box 12277, Research Triangle Park, N.C. 27709.

www.isa.org

- S12.12-1995, Electrical Equipment for Use in Class 1, Division 2, Hazardous (Classified) Locations.

- S 71.01-1985, Environmental Conditions for Process Measurement and Control Systems: Temperature and Humidity.

National Fire Protection Association (NFPA)

One Batterymarch Park

P.O. Box 9101, Quincy, Mass. 02269

www.nfpa.org

- NFPA 1-1992, Fire Prevention Code.

- NFPA 15-1990, Water Spray Fixed Systems for Fire Protection.

- NFPA 16-1995, Installation of Deluge Foam-Water Sprinkler and Foam-Water Spray Systems.

- NFPA 30-1993, Flammable and Combustible Liquids Code.

- NFPA 49-1991, Hazardous Chemicals Data.

- NFPA 70-1990, National Electrical Code.

  NFPA 704-1990, Identification of the Fire Hazards of Materials.

Figure 8-7 Consensus Organizations and Standards (continued)

Premium rates will be affected if management does not comply or negotiate corrective action for the items on the inspector's exception report.

Consensus standards are more comprehensive and detailed than OSHA standards. (OSHA standards such as HAZCOM and PSM define the *minimum* requirements for employers to follow in order to assure chemical safety and health.) Because consensus standards teach users how to meet and exceed regulatory standards, they are an excellent source for defining facility policy and procedures. Most designers turn to consensus standards such as NFPA for specifications and other design information. Contact ANSI or the other organizations directly to find standards not listed in Figure 8-3 or the OSHA code.

Environmental Protection Agency

Environmental Protection Agency (EPA) requirements are often considerably more comprehensive and complex than those provided by OSHA. For example, consider the scope of the eight major statutes directly enforced by the EPA:

- The Solid Waste Disposal Act; amended by the Resource Conservation and Recovery Act (RCRA)
- The Toxic Substances Control Act (TSCA)
- The Comprehensive Environmental Response, Compensation, and Liability Act (CERCLA or Superfund; includes EPCRA, or Emergency Planning and Community Right to Know Act, as Title III of the Superfund Amendments and Reauthorization Act of 1986 (SARA))
- The Clean Air Act (CAA)
- The Federal Water Pollution Control Act; also known as the Clean Water Act (CWA);
- The Federal Insecticide, Fungicide, and Rodenticide Act (FIFRA);
- The Marine Protection, Research, and Sanctuaries Act (MPRSA);
- The Safe Drinking Water Act (SDWA)

Detailed information regarding these statutes may be found from the EPA home page, www.epa.gov.

EPA Standards for Chemical Safety Management

The first four statutes are most relevant to the treatment of chemical safety management.[11]

The Resource Conservation and Recovery Act (RCRA)

RCRA, originally passed in 1976, provides a comprehensive federal program for regulating solid wastes and especially hazardous waste. RCRA is the so-called "cradle to grave" control of hazardous waste.

In 1984, amendments were made in order to tighten restrictions on waste recycling; land disposal operations; and contaminant releases from old, abandoned, and contaminated sites not covered by the original RCRA. The 1984 amendments also changed requirements for small quantity hazardous generators and established a new program for leaking underground storage tanks. An EPA call center has been established primarily to answer questions about RCRA, Superfund, and EPCRA.

Toxic Substances Control Act (TSCA)

The Toxic Substances Control Act of 1976 deals with the manufacture and use of chemical products, including polychlorinated biphenyl (PCB) materials, commonly found in old transformers. The detailed regulations cover use, management, storage, and disposal of PCBs and other chemicals, and establish complex requirements for inspection, recordkeeping, reporting, and testing. Chemical manufacturers, distributors, and processors are included under TSCA. These operations take into account chemical importing and exporting, as well as chemicals included in hazardous waste. Covered facilities should consult an environmental regulatory expert to determine whether their activities are in compliance with TSCA.

Comprehensive Environmental Response Compensation and Liability Act

CERCLA or Superfund, passed in 1980, provides the authority and funding for cleaning up (1) spills and discharges of hazardous substances into the environment and (2) inactive hazardous waste dumpsites.

The Superfund Act requires hazardous material sites to notify the EPA when a non-permitted spill or discharge occurs in order to facilitate quick cleanup. Owners of inactive HAZMAT sites are to notify EPA of any known disposal or discharge of material, including the amount and type of material. EPA may order remedial action by *potentially responsible parties* (PRPs), including the current owner and operator of the site; the owner and operator of the site at the time of HAZMAT disposal; any party that transported hazardous material to the disposal or treatment site at its own choosing; and any party that arranged for the disposal or treatment of the material. The Superfund Amendments and Reauthorization Act (SARA) of 1986 added significant changes to the original act meant to accelerate cleanup, expand public participation, encourage voluntary settlement, and make other provisions.

Emergency Planning and Community Right-to-Know (EPCRA)

The Community Right-to-Know Act was an independent part of Title III of SARA . The two-part intent was:

1. To ensure a way for concerned citizens to be aware of hazardous materials that could impact their communities
2. To require state and local governments to develop emergency response plans involving these materials

Each state was required to establish a state emergency response commission (SERC) that was responsible to designate Local Emergency Planning Committees (LEPCs). Major provisions of the law included emergency planning, emergency release notification, material safety data sheet (MSDS) reporting requirements, hazardous chemical inventory reporting, toxic chemical release reporting (Form R), and supplier notification.

Clean Air Act (CAA) and Amendments

The Clean Air Act, passed in 1970, mandates pollution control requirements for sources of air pollution. Federal facilities may have regulatory responsibilities under the Clean Air Act including obtaining emission permits, installing pollution and emission control devices, developing risk management plans, and maintaining records. Part of EPA's mission is to ensure that Federal facilities comply with these requirements.[12] The Clean Air Act Amendments (CAAA) of 1970, 1977, and 1990 have added significantly to the original statute. For example, the CAA amendments of 1990 significantly expanded EPA requirements for mobile and stationary sources (including manufacturers, processors, refiners, and utilities). These facilities or stationary sources are now required to obtain operating permits that limit air emissions. EPA speaks of five specific themes accomplished by the CAA amendments:[13]

- "Encourage the use of market-based principles and other innovative approaches, like performance-based standards and emission banking and trading;
- Provide a framework from which alternative clean fuels will be used by setting standards in the fleet and California pilot program that can be met by the most cost-effective combination of fuels and technology;
- Promote the use of clean low sulfur coal and natural gas, as well as innovative technologies to clean high sulfur coal through the acid rain program;
- Reduce enough energy waste and create enough of a market for clean fuels derived from grain and natural gas to cut dependency on oil imports by one million barrels/day; and
- Promote energy conservation through an acid rain program that gives utilities flexibility to obtain needed emission reductions through programs that encourage customers to conserve energy."

EPA's comprehensive RMP performance standard is an example of the first theme. Here, many facilities have to conform to a new rule to protect the public and the environment from accidental releases of hazardous materials: 40 CFR 68—Accidental Release Prevention Provisions, Risk Management Planning (RMP). The RMP regulation was published by EPA June 20, 1996 under the CAA section 112(r)(7) and finalized in 1999. Facilities were given three years to submit risk management plans to EPA. This rule will be discussed in more detail below.

The CAAA of 1990, Title V, also brought far-reaching measures for industrial facilities. Section 501 established a new national operating permit program similar to that for the Clean Water Act. Covered emission sources have to obtain an operating permit that requires renewal at least every 5 years and pay an annual fee.

Enforcement Actions

EPA refers civil and criminal enforcement actions to the U.S. Department of Justice (DOJ). For a number of reasons, DOJ has historically been more aggressive in enforcing EPA statutes within private industry than in prosecuting "sister" agencies, but that is changing, and now man-

Table 8-4 EPA Enforcement Actions Taken Under Statutes, 1972-1997[14]

Year	CAA	CWA/ SDWA	RCRA	CERCLA	FIFRA	TSCA	EPCRA	Total
1972	0	0	-	-	880	-	-	880
1973	0	0	-	-	1274	-	-	1274
1974	0	0	-	-	1387	-	-	1387
1975	0	738	-	-	1614	-	-	2352
1976	210	915	0	-	2488	0	-	3613
1977	297	1128	0	-	1219	0	-	2644
1978	129	730	0	-	762	1	-	1622
1979	404	506	0	-	253	22	-	1185
1980	86	569	0	0	176	70	-	901
1981	112	562	159	0	154	120	-	1107
1982	21	329	237	0	176	101	-	864
1983	41	781	436	0	296	294	-	1848
1984	141	1644	554	137	272	376	-	3124
1985	122	1031	327	160	236	733	-	2609
1986	143	990	235	139	338	781	0	2626
1987	191	1214	243	135	360	1051	0	3194
1988	224	1345	309	224	376	607	0	3085
1989	336	2146	453	220	443	538	0	4136
1990	249	1780	366	270	402	531	206	3804
1991	137	1745	364	269	300	422	179	3416
1992	354	1977	291	245	311	355	134	3667
1993	279	2216	282	260	233	319	219	3808
1994	435	1841	115	264	249	333	307	3544
1995	232	1774	92	280	160	187	244	2969
1996	242	998	238	334	83	178	198	2171
1997	391	1642	423	305	181	185	300	3427

agers of federal facilities are focusing on prevention, preparedness, and response activities as they relate to chemical safety management.

In the past, EPA has focused most of its enforcement activities on sources of air, water, or land pollution. Recently, however, EPA has moved toward multimedia enforcement, or filing actions against a single facility under several different statutes. Table 8-4 presents the number of enforcement actions taken under the statute acronyms above from 1972 through 1997.

List of Lists

Using EPA standards, it is often a challenge to determine all of the regulated chemicals that may be processed, generated, or used at a facility. To help, EPA has recently updated the Consolidated List of Chemicals Subject to the EPCRA and RMP regulations, also known as the list of lists. The list should be used only as a reference tool, not as a source of compliance information. The information is listed alphabetically and by chemical abstract number (CAS #). Users can download the list of lists (as well as other current EPA information) starting from the EPA web page.

Risk Management Program Rule and OSHA PSM

Much of EPA's RMP rule contains similar provisions to OSHA's regulation. EPA essentially adopts PSM as the prevention program for accidental chemical release. However, there are also important differences, such as the list of covered chemicals, three process program levels introduced by EPA, the RMP hazard assessment, certain elements that differ between RMP and PSM, etc. We will compare the two regulations, point out important differences, and integrate requirements wherever possible in order to make implementation more straightforward.

There are three major components to the RMP rule:

1. A hazard assessment to estimate potential release quantities of toxic and flammable materials and to estimate downwind effects, including the number of people that could be harmed. The assessment must include a history of releases over the past five years and an evaluation of a worst-case release scenario.

2. A prevention program for preventing accidental releases (nearly identical in many elements to PSM).

3. An emergency response plan including provisions for notifying the public and local first-responders, emergency health care services, and employee training.

Figure 8-8 is an outline of the RMP rule.

The plan itself is under Subpart G. The Hazard Assessment, Prevention Plans, and Emergency Response sections are found under subparts B, C/D, and E, respectively. Subpart F (68.130) lists the chemicals and a table of toxic endpoints are found in Appendix A.

Subpart A- General
68.1   Scope
68.2   Stayed provisions
68.3   Definitions
68.10   Applicability
68.12   General requirements
68.15   Management

Subpart B – Hazard Assessment
68.20   Applicability
68.22   Offsite consequence analysis parameters
68.25   Worst-case release scenario analysis
68.28   Alternative release scenario analysis
68.30   Defining offsite impacts – population
68.33   Defining offsite impacts – environment
68.36   Review and update
68.39   Documentation
68.42   Five-year accident history

Subpart C – Program 2 Prevention Program
68.48   Safety information
68.50   Hazard review
68.52   Operating procedures
68.54   Training
68.56   Maintenance
68.58   Compliance audits
68.60   Incident investigation

Subpart D – Program 3 Prevention Program
68.65   Process safety information
68.67   Process hazard analysis
68.69   Operating procedures
68.71   Training
68.73   Mechanical integrity
68.75   Management of change
68.77   Pre-startup review
68.79   Compliance audits
68.81   Incident investigation
68.83   Employee participation
68.85   Hot work permit
68.87   Contractors

Subpart E – Emergency Response
68.90   Applicability
68.95   Emergency response program

Subpart F – Regulated Substances for Accidental Release Prevention
68.100   Purpose
68.115   Threshold determination
68.120   Petition process
68.125   Exemptions
68.130   List of substances

Figure 8-8 EPA Accidental Release Prevention Provisions

Subpart G – Risk Management Plan
68.150   Submission
68.151   Assertion of claims of confidential business information
68.152   Substantiating claims of confidential business information
68.155   Executive summary
68.160   Registration
68.165   Offsite consequence analysis
68.168   Five-year accident history
68.170   Prevention program/program 2
68.175   Prevention program/program 3
68.180   Emergency response program
68.185   Certification
68.190   Updates

Subpart H – Other Requirements
68.200   Record keeping
58.210   Availability of information to the public
68.215   Permit content and air permitting authority or designated agency requirements
68.220   Audits
APPENDIX A TO PART 68 – TABLE OF TOXIC ENDPOINTS
AUTHORITY: 42 U.S.C. 7412(r). 7601(a)(1), 7661-7661f
SOURCE: FR Doc. 99-231 Filed 1-5-99.

Figure 8-8 EPA Accidental Release Prevention Provisions (continued)

Covered Industries

The OSHA PSM standard is aimed primarily at the chemical and petroleum industries where there has been a history of accidents. EPA covers 10 industries according to the North American Industrial Classification System (NAICS) as Program Level 3, as well as processes covered by OSHA under PSM (see Figure 8-9). This means they must implement OSHA PSM requirements as the RMP prevention program.

	Industry	NAICS
1.	Pulp Mills	322110
2.	Petroleum Refineries	324110
3.	Petrochemical Manufacturing	32511
4.	Alkalis And Chlorine Manufacturing	325181
5.	All Other Basic Inorganic Chemical Manufacturing	325188
6.	Other, Cyclic Crude And Intermediate Manufacturing	325192
7.	All Basic Organic Chemical Manufacturing	325199
8.	Plastic Material And Resin Manufacturing	325211
9.	Nitrogenous Fertilizer Manufacturing	325311
10.	Pesticide And Other Agricultural Chemical Manufacturing	325320

Figure 8-9 RMP Program 3 Processes

Some manufacturing sectors without a history of reported releases, such as ethyl alcohol manufacturing from corn and other grains (NAICS 325193) are only covered under RMP if they utilize other listed toxic or flammable materials. Ethanol plants are covered under the OSHA PSM regulation if they are over the 10,000 lb. threshold for flammable liquids or gases.[15]

A full list of the new NAICS codes and further information is available from the U.S. Bureau of the Census.[16]

General Duty Clause

The EPA general duty clause prevents facilities from "opting-out" of RMP based on a technicality such as having a quantity of hazardous chemical just under the threshold quantity. No accidents may occur for years, but in the event of a release, the offending facility is likely to be prosecuted under the general duty clause.

Owner/Operators of sources...
"... producing, processing, handling or storing..."
"...any substance listed ... or any other extremely hazardous substance"
Have a general duty to:
    identify hazards, which may result from releases;
    design and maintain a safe facility; and
    minimize the consequences of releases, which do occur.

Figure 8-10 EPA General Duty Clause[17]

From an enforcement standpoint, EPA considers the requirements of RMP (Figure 8-8) and the general duty clause to be one and the same.

RMP Program Levels

All PSM facilities must meet the same requirements. However, EPA has separated eligibility criteria into three levels, depending on accident history, facility location relative to public receptors, and other factors:

RMP Program 1

No accidental release during the last five years with severe offsite consequences; worst-case endpoint is less than the distance to any public receptor; and the facility has coordinated emergency response procedures with local first responders. (Program 1 processes have the lowest level of requirements.)

RMP Program 2

Program 2 covers all processes that do not fall into either Program 1 or Program 3 classification.

RMP Program 3

The covered process is in one of ten paper, chemical, or petroleum classification codes or is subject to the requirements of OSHA Process Safety Management. See Figure 8-9 above.

The safety and health manager is well advised not to split hairs in order to classify itself under Program 1 or 2, but to fully implement a Program 3 prevention plan that meets all PSM requirements.

Comparison of Chemical Lists

There are 140 chemicals on the EPA list for accidental release prevention requirements, either toxic (77 chemicals) or flammable gases and liquids (63 substances). Congress mandated an initial list of sixteen specific substances as Hazardous Air Pollutants (HAPs) under Section 112 of the Clean Air Act Amendments because there had been a history of accidental releases involving these materials:

- Ammonia
- Anhydrous ammonia
- Anhydrous hydrogen chloride
- Anhydrous sulfur dioxide
- Bromine
- Chlorine
- Ethylene oxide
- Hydrogen cyanide
- Hydrogen fluoride
- Hydrogen sulfide
- Methyl chloride
- Methyl isocynanate
- Phosgene
- Sulfur trioxide
- Toluene diisocyanate
- Vinyl chloride

There are 24 RMP listed materials and threshold quantities that are not covered by OSHA (15 are flammable; 9 are toxic, indicated by (T)):

1.	Acrylonitrile	20,000 lbs
2.	Allyl alcohol	15,000 lbs
3.	(T)Arsenous trichloride	15,000 lbs
4.	(T)Boron trifluoride compound with methyl ether	15,000 lbs
5.	Carbon disulfide	20,000 lbs
6.	(T)Chloroform	20,000 lbs
7.	Crotonaldehyde	20,000 lbs
8.	Cyclohexylamine	15,000 lbs
9.	Epichlorohydrin	20,000 lbs
10.	Ethylenediamine	20,000 lbs
11.	(T)Hydrazine	15,000 lbs
12.	Hydrochloric acid	15,000 lbs
13.	Isobutyronitrile	20,000 lbs
14.	(T)Isopropyl chloroformate	15,000 lbs
15.	(T)Methyl thiocyanate	20,000 lbs
16.	Piperidine	15,000 lbs
17.	Propionitrile	10,000 lbs
18.	Propyleneimine	10,000 lbs
19.	Propylene oxide	10,000 lbs
20.	(T)Tetranitromethane	10,000 lbs
21.	(T)Titanium tetrachloride	2,500 lbs
22.	(T)Toluene diisocyanate	10,000 lbs
23.	Trimethylchlorosilane	10,000 lbs
24.	Vinyl acetate monomer	15,000 lbs

The EPA chemical list is contained in four tables and a chemical may be listed for any of the following reasons:

a) Mandated by Congress
b) On EHS list, vapor pressure 10 mm Hg or greater
c) Toxic gas
d) Toxicity of hydrogen chloride, potential to release, and history of accidents
e) Toxicity of sulfur trioxide and sulfuric acid, potential to release sulfur trioxide, and history of accidents
f) Flammable gas
g) Volatile flammable liquid

Table 8-5 Partial Chemical List Covered by OSHA and EPA

| Chemical | Threshold Quantity | | OSHA | EPA RMP[4] |
	OSHA PSM[1]	EPA RMP (lbs)[2]	HAZCOM [3] PEL, Mg/m³	Toxic End-Point Mg/m³
*Acetaldehyde	2500	10,000	360	flammable
*Acrylonitrile	-	20,000	4.6	76
Anhydrous ammonia	10,000	10,000	35	140
Bromine	1500	10,000	0.7	6.5
*Butadiene	-	10,000	2200	flammable
*Chlorine	1500	2500	C3	8.7
*Chloroform	-	20,000	C240	490
Dimethylamine, anhydrous	-	10,000	18	flammable
*Dimethylhydrazine	1000	15,000	1	12
Ethylamine	7500	10,000	18	flammable
*Ethylene oxide	5000	10,000	9	90
Fluorine	1000	1000	0.2	3.9
*Formaldehyde	1000	15,000	1.2	12
*Hydrogen chloride	5000	5000	C7	30
*Hydrogen fluoride	1000	1,000	2	16
Isopropyl amine	5000	10,000	5	flammable
Methylamine	1000	10,000	12	flammable
*Methyl Hydrazine	100	15,000	C.35	9.4
Nitric Acid	500	15,000	5	26
Nitric oxide	250	10,000	30	31
Oxygen difluoride	100	-	0.1	
Ozone	100	-	0.2	
Perchloryl fluoride	5000	-	13.5	
*Phosgene	100	500	0.4	.81
Sulfur dioxide	1000	5000	13	.78
Sulfuric acid (oleum)	1000	10,000	1	10
*Toluene Diisocynanate	-	10,000	C14	7
*Vinyl acetate monomer	-	15,000	-	260
*Vinyl chloride	-	10,000	12.6	flammable

* Hazardous Air Pollutant under Section 112 of the Clean Air Act Amendments

(1) Threshold quantity in pounds to be covered by OSHA Standard CFR 1910.119

(2) Threshold quantity in pounds to be covered by EPA 40 CFR Part 68

(3) 8-hour permissible exposure limit (PEL), time weighted average. C = ceiling limit, exposure not to exceed limit at any time

(4) Toxic end point listed in Appendix A of Part 68 and converted to milligrams/cubic meter.

EPA threshold quantities (TQs) are greater than or equal to OSHA's TQs except for methyl chloride (EPA TQ = 10,000 lbs., OSHA TQ = 15,000 lbs.).

OSHA covers flammable liquids and gases (without specific listing) in quantities of 10,000 pounds, except for hydrocarbon fuels, such as propane, which are used solely for workplace consumption as a fuel. EPA coverage of propane, butane, ethane, methane, and other fuels has recently been dropped.

Table 8-5 provides a partial list of chemicals covered by both EPA and OSHA.

In most cases, the OSHA permissible 8 hour exposure limit (PEL) is lower than EPA's toxic endpoint limit as listed in Appendix A of the RMP standard.

RMP Hazard Assessment

The RMP rule requires facilities to conduct a hazard assessment for all covered processes; this is not a provision of the PSM standard. These assessments require an off-site consequence analysis or worst-case release scenarios (for program level 2 and 3 processes) and documentation of a five-year accident history involving RMP-regulated materials. EPA has published a guidance document (RMP Offsite Consequence Analysis Guidance) available through the home page of EPA's Chemical Emergency Preparedness and Prevention Office that provides guidance to help facilities perform the studies (see Chapter 1 and Appendix A and B).

The first step in hazard assessment is to determine the worst-case release scenario, including:

- Rationale for selecting the scenario
- Estimated quantity released
- Release rate and duration of the release
- Distance to a toxic or flammable endpoint (endpoint specified by EPA for each material, see Table 8-5)
- Estimated population and environmental receptors affected

The worst-case quantity released gives the greatest distance to a toxic or flammable endpoint. Administrative controls and passive mitigation systems in place may be used to reduce the offsite impact. Note that for program level 1 processes, facilities must study at least one worst-case scenario for toxicants as a class and one worst-case scenario for flammables.

Alternate release scenarios (those more likely to occur than the worst-case scenario) must be analyzed and reported for program level 2 and 3 processes. The alternative release scenario travel distance to the endpoint must extend off-site. Off-site consequence analysis must be reviewed and updated once every five years unless changes are made to a process that increases or decreases the offsite results by a factor of two or more. In that event, the off-site consequence analysis must be revised and resubmitted in the risk management plan within six months.

The hazard assessment also requires facilities to conduct and document a five-year accident history and note incidents that had significant on or off-site consequences. For OSHA-covered processes, these accidents should be documented as part of the PSM incident investigation program.

Prevention Program

By implementing level 3 requirements, a facility assures that they have a single prevention program that meets both EPA and OSHA regulations.

Program level 2 prevention requirements are comparable but less stringent than the corresponding PSM elements. For example, the RMP rule does not require use of a team to conduct a hazard review and allows use of a checklist.[17] The facility process safety coordinator should use extreme caution when tempted to take shortcuts in order to complete the prevention plan.

Program 1 facilities do not need a formal management system. No alternative release scenario is required of the hazard assessment. No prevention program is needed, but the facility is still subject to the EPA general duty clause.

Emergency Response

The main difference between EPA and OSHA with respect to emergency response is the RMP requirement to coordinate the facility plan with the community emergency response plan. Facilities must inform the local emergency response agency of any information affecting implementation of the community plan. EPA also requires procedures for the use, inspection, testing, and maintenance of emergency response equipment.

Implementing the Risk Management Program

The steps to installing a RMP program are:

1. Identify all OSHA and EPA covered materials used on site.
2. Identify all EPA covered processes.
3. Identify processes also covered under OSHA PSM for which a PHA has been completed.
4. Schedule and conduct PHA(s) on those covered by EPA where no PHA has yet been done. Consider the five-year accident history and deviations with potential offsite consequences.
5. Begin implementation of other Process Safety Management elements. Integrate PSM with HAZCOM and other relevant OSHA programs.
6. Select one worst-case scenario for toxicants and one worst-case scenario for flammables. Complete worst-case and alternate scenarios using RMPComp (consider other models).
7. Develop at least one alternative release scenario for each toxic chemical and one to represent all flammables with potential offsite consequences.
8. Begin communication planning with employees, the community, and other stakeholders.
9. Begin compiling information using the RMP Data Elements check list.
10. Complete draft of RMP Plan and send to the RMP Reporting Center using RMP Submit.
11. Review the plan with the local emergency planning committee (LEPC) or equivalent.

Communication Issue

One issue that often comes up is "How do we design the message for employees and the public?

Think of your prevention and EP&R programs as weights on either end of a balance. Develop balanced, integrated communication plans for employees and the community.

Consistent with community right-to-know requirements, EPA has provided free software (RMP*Submit™) for facilities to use in submitting RMPs. The public was to have internet access to this information, including offsite consequence analysis, via a program called RMP*Info™. Furthermore, facilities in Program 2 or 3 categories were required to hold a public meeting to discuss the RMP. EPA said, "The information is intended to stimulate the dialogue between industry, state and local officials, and the public to improve accident prevention and emergency response practices."

Pressure from stakeholders prompted a study to determine the risk of terrorism due to posting sensitive information on the internet.[18] "The study found that if the information were to be posted, the risk of terrorist attack would rise by 250%. The study called this increase incremental because of the low probability of an event's occurring in the first place."[19]

Due to increased concern for public safety and chemical site security following September 11, 2001, EPA has temporarily removed RMP*Info from its website. Check the website before developing your external communications plan (Step 7, Chapter 10).

Information and Resources

There is much current and useful information available on the EPA/CEPPO website. See links to the following references:

- CEPPO's RMP Guidance by Industry Sector
- Center for Chemical Process Safety (CCPS) of the American Institute for Chemical Engineers (AIChE)
- American Petroleum Institute (API)
- National Association of Chemical Distributors (NACD)
- The American Chemistry Council (formerly the Chemical Manufacturers Association)

Small business resources include:

- RMP Network: South Carolina Small Business Assistance Program
- Small Business Assistance Program
- Enviro$ense: Small Business Assistance
- Guidelines for Process Safety in Outsourced Manufacturing Operations

Department of Transportation Standards

Facility safety coordinators need a basic understanding of DOT standards to integrate transportation and emergency response requirements for covered chemicals into plant policies and procedures. Relevant DOT standards are grouped within the general category of HAZMAT safety.

DOT oil spill prevention and response plans plus hazardous materials (HAZMAT) regulations are found in 49 CFR Parts 130-180. DOT's chemical list (Hazardous Materials Table), special provisions, HAZMAT communications, emergency response information, and training requirements are all found in 49 CFR Part 172. The subparts A-H provide the following specific requirements:

A. General—purpose, scope, and general applicability
B. Table of Hazardous Materials and special provisions
C. Shipping Papers—descriptions required for shipping papers, shipper's certification, and hazardous waste manifest
D. Marking—general identification marking requirements and coverage for radioactive, liquid, poisonous, explosive, marine pollutants, high temperature materials, and HAZMAT materials in portable tanks cargo tanks and tank cars
E. Labeling—labeling requirements for explosives, non-flammable gas, flammable gas, flammable liquids and solids, materials that are spontaneously combustible, dangerous when wet, oxidizers, poisonous, infectious, radioactive, and corrosive
F. Placarding—placarding requirements for HAZMAT materials
G. Emergency response information
H. Training—training for people in the shipping chain including operator's cargo staff; employees who handle, store, and load dangerous goods; passenger-handling; operator and security staff; flight and other crew members; packers, shippers, and agents

Hazardous Materials Table

The Hazardous Materials Table lists shipping requirements for thousands of regulated chemicals, including:

• Aerosol products, paints, and adhesives
• Pesticides that include solids, liquids, and gases
• Paint thinners, degreasers, and other solvent products

- Soil core samples potentially contaminated with volatile organic compounds or other regulated materials
- Chemicals including acids, bases, and alcohols

The table provides critical information for shippers that includes what regulated materials can be shipped and how, maximum quantities allowed, correct shipping papers and labels, emergency response information, and required training for hazmat personnel.

If a facility is to avoid a substantial penalty, plant management and the safety, health, and environmental coordinator(s) must be sure all transportation and logistics people understand and follow these detailed requirements.

DOT coordinators need to have a printed version of the 49 CFR manual for daily reference. In general, emergency response information required with a HAZMAT shipment includes:

1) Basic description and technical name
2) Immediate hazards to health
3) Risks of fire or explosion
4) Immediate precautions to be taken in the event of an accident or incident
5) Immediate methods for handling fires
6) Initial methods for handling spills or leaks in the absence of fire
7) Preliminary first aid measures

HAZMAT Safety Home Page

Current DOT chemical safety information can be found on the web at www.hazmat.dot.gov.

The reader can also find information on HAZMAT enforcement, spills, state plans, markings, labels, emergency response, training, and other topics.

Emergency Response Guidebook

Another important reference for site safety and health coordinators is the Emergency Response Guidebook. The ERG was developed by the U.S. DOT, Transport Canada, and the Secretariat of Transport and Communications of Mexico. It is designed for fire fighters, police, and other emergency responders arriving on the scene of a transportation incident involving HAZMAT materials. Because the guide illustrates the DOT hazard classification system and presents other information in a simple format, it is extremely useful for establishing chemical safety policies and procedures that meet or exceed DOT requirements.

The guide features four color-coded sections to aid emergency personnel with quick identification of specific and generic material hazards, and for protecting themselves, the general public, and others during the initial response phase of the incident. The four sections are:

*   Yellow—an index list of dangerous goods in numerical order of ID number;
*   Blue—an index of dangerous goods in alphabetical order of material name;
*   Orange—Safety recommendations and emergency response information listed in order of guide number. Sections here are potential material hazards, suggested public safety measures, and emergency response actions, including first aid.
*   Green—Table of :
    1. Initial isolation
    2. Table of initial isolation and protective action distances

Two distances are provided for small and large spills and are further divided into day and night responses.

Other features of the guide book are examples of shipping papers and placards, safety precautions when approaching a transportation incident, instructions on who to call for assistance, the Class 1-9 DOT hazard classification system, a table of placards and initial response guides to use on-scene, and rail car and road trailer identification charts.

The ERG is a must for safety, health, and environmental coordinators. It is distributed free of charge to public safety organizations and may not be resold. The guidebook is normally revised and reissued every three or four years. Users should check every six months or so to be sure their version is current. Contact: http://hazmat.dot.gov/gydebook.htm or http://www.tc.gc.ca/canutec/en/guide/guide-e.htm in Canada.

Example 1—Benzene

The Table of Hazardous Materials requirements for shipping benzene includes:

Description, shipping name: Benzene

Identification Numbers: UN1114

Packing Group: II

Label Codes: 3 (Flammable and combustible liquids)

Passenger aircraft/rail: 5L (5 liters max.)

Cargo aircraft only: 60L

To locate information on benzene in the ERG, look up benzene alphabetically in the blue section. There the guide number 130 is given. Under guide 130 in the orange section are potential fire/explosion and health hazards, public safety information, and what to do in the event of a fire, spill, or leak for non-polar materials such as benzene. First aid procedures are also included here.

Example 2—Denatured Alcohol

DOT classifies the material as a class 3 flammable liquid. The DOT ID number is NA 1987. The shipping limitations are 5 liters and 60 liters for passenger aircraft/rail and cargo aircraft, respectively.

From the ERG, the guide number for 1987 is 127. Although the shipping limitations are the same as benzene, the emergency response instructions for this chemical are for a polar/water-miscible flammable liquid—very different from the information given for benzene.

Example 3—Anhydrous Ammonia

The UN number is 1005 and DOT informs the reader that anhydrous ammonia must be shipped in pressurized containers with pressure relief safety devices. Cargo planes only may carry a maximum of 25 kg in one package. The warning "Inhalation Hazard" must be on shipping papers and container placards.

The ERG guide number for ammonia is 125, corrosive gases. We are reminded that the consequences of inhalation may be fatal, consistent with DOT shipping instructions and the MSDS information. The ERG goes on to list health, fire, or explosion hazards for corrosive gases and what to do in terms of public safety and emergency response.

Example 4—Carbon Dioxide

The product is listed in the DOT Hazardous Material Table as UN # 1013, hazard class 2.2 (for non-flammable, non-toxic compressed gases). Shipping limitations are 75 kg (passenger aircraft/rail) and 150 kg (cargo aircraft only).

The ERG guide number for carbon dioxide is 120, inert gases, including refrigerated liquids. The main health hazards for emergency responders are sudden dizziness or asphyxiation if inhaled and frostbite/tissue burns if exposed to the liquefied gas. Most know that carbon dioxide does not burn; however ruptured cylinders may rocket if an external fire causes a release.

Summary

In summary, facility safety coordinators must work with shipping and distribution departments to integrate DOT, EPA, and OSHA requirements into site policies and procedures. This means training for the potential hazards of materials on-site, on the road, and for protection of public safety. In addition, plant and off-site emergency responders need to know what to do in the case of a fire, spill, or leak and how to provide first aid.

Integrating Standards into Policy

An important step in the continuous improvement of safety is moving from regulatory compliance to implementation of company policy (see Chapter 10).

Figure 8-11 depicts a typical structure of codes and standards.

At the top of the structure as company and facility policies and procedures; consensus or teaching standards are at the base. Federal, state, and local codes and standards are shown in the middle. Local codes and standards are above federal and state standards because many federal standards contain the phrase "or according to the authority having jurisdiction." In other words, state and local codes are usually written to meet or exceed federal standards. For example, requirements for protecting people and property from flammable liquids and gases may come from a city building code or be based on the Universal Building Code (UBC). In the event of a fire, the means of egress (path of escape) is specified by federal OSHA and references the consensus standard NFPA 101, Life Safety Code.

**COMPANY AND FACILITY
POLICIES AND PROCEDURES**

- Safety, Health and Environmental Policy
- Hazard Communication
- Process Safety Management
- Workplace Inspection Program
- Safety and Health Training
- Lockout-Tagout Procedures
- Confined Space Entry Procedures

LOCAL CODES AND STANDARDS

- City Building Codes
- Local Fire Codes
- Local Standards for Public Safety

FEDERAL AND STATE CODES AND STANDARDS

- OSHA Hazard Communication
- OSHA Process Safety Management
- OSHA Hazardous Waste Operations and Emergency Response
- EPA Prevention of Accidental Releases (Risk Management Planning)
- EPA Hazardous Air Pollutants
- DOT Hazardous Materials Regulations and Procedures

NATIONAL CONSENSUS STANDARDS

- American National Standards Institute (ANSI)
  - ❑ K61.1 Storage and Handling of anhydrous ammonia
- American Petroleum Institute
- American Society of Mechanical Engineers (ASME)
  - ❑ ASME boiler and pressure vessel code, section VIII
- American Society for Testing and Materials
- Chlorine Institute
- National Fire Protection Association (NFPA)
  - ❑ NFPA 30 Flammable and combustible liquids code

Figure 8-11 Structure of Safety and Health Codes and Standards

Write company or facility policy to meet or exceed federal, state, and local requirements. To assure accuracy and save time, use the same outline as the regulatory standard. For example, when writing HAZCOM policy, start with Figure 8-2 and use the categories a)-i). Describe the OSHA requirements in each category using your facility's language and add a section called comments for a detailed explanation of policy and procedures.

Example 1

Several years ago, a plant manager expressed the facility's confusion over the parade of environmental, health, and safety (EHS) reviews coming through the facility each year. He said, "After this, I want all the requirements and what we're doing about them in my war room!" The war room was where product and process specifications were kept and so it made sense to place OSHA, EPA, and DOT code books there too. Facility EHS standards were added as they were written and implemented.

Example 2

The author has helped plants establish safety libraries containing OSHA, EPA, and DOT codebooks, as well as the National Safety Council's three-volume Accident Prevention Manual and other EHS references. File drawers were set up with OSHA code files in front (e.g. subparts A-Z, Table 8-3) and plant program files behind each OSHA category. This type of integration helps safety coordinators, plant management, and employee action teams understand and communicate their safety programs. A further benefit occurs when the OSHA inspector comes to call. Regardless of who the plant representative is, he or she can bring the inspector to one place rather than search the facility to find the documentation needed to demonstrate compliance.

Conclusion

The important personnel-type OSHA regulations relative to chemicals are found in subpart H (hazardous materials) except for 29 CFR 1910.1200 Hazard Communication (HAZCOM). The process safety management regulation, 29 CCFR 1910.119 is a comprehensive performance-based standard that applies key components of HAZCOM to prevention of catastrophic process related incidents.

Chemical safety managers should consider purchasing written safety policy templates for chemical labels, hazard communication, process safety, MSDS development program, etc. that help the organization customize policies and procedures. It is critical that the final standard states "who, what, where, and how a procedure is to be done" at the facility and does not simply a copy, cut, and paste paper exercise.[20]

EPA regulations most related to chemical safety management are the Resource Conservation and Recovery Act (RCRA), Toxic Substances Control Act (TSCA), Comprehensive Environmental Response Compensation and Liability Act (CERCLA or Superfund), the Emergency Planning, Community Right-to-Know Act (EPCRA), and the Clean Air Act (CAA) and Amendments. EPA's risk management regulation related to OSHA's PSM is based on the CAA amendments. Here the EPA rule is meant to protect people and environmental assets that are offsite. All OSHA regulations, including PSM, are concerned with on-site exposures.

Relevant chemical transportation standards are available from DOT's Office of Hazardous Materials (HAZMAT) Safety homepage. The Emergency Response Guidebook (ERG) is a good reference source for facility safety, health, and environmental coordinators. Use the ERG to quickly identify general and specific HAZMATs by number and for training of emergency response personnel.

Chemical safety managers need to be guided by regulatory standards when first developing their safety infrastructure. Note that chemical-related facilities must comply with all regulations associated with any facility plus those directed at controlling releases and exposures to specific hazardous materials. Start with the facility's chemical list and search OSHA, EPA, and DOT regulations for specific requirements.

While it is important to adopt the regulatory requirements initially, a program will stagnate if stopped there. Organizations must go beyond regulatory compliance to conforming to related company and facility policies and procedures. A vision needs to be in place before developing safety policies and procedures. Establish accountability and measurements to drive continuous improvement (Chapter 7). Then regulatory standards serve as an excellent guide for drafting policies, procedures, and design/purchase standards. Use OSHA's alpha-numeric system and titles and add company and/or facility-specific comments to customize internal standards.

A common mistake production or maintenance workers (and some safety coordinators) make is to get into a "where is it written" mentality. That is, people look for an OSHA regulation or workplace rule that addresses the safe/unsafe work condition or procedure. If the regulation, policy, or procedure goes against current or proposed practice, the argument is "tell me *why* this is or isn't safe!" Readers will do well not to engage in that argument but rather brainstorm with the supervisor/manager/worker "what is the hazard?" That is the true value of regulatory requirements— they address real or potential hazards.

References

1. *Plant Guidelines for Technical Management of Chemical Process Safety*, CCPS of the American Institute of Chemical Engineers, 1992, p. 254.

2. "Henshaw's Brave New World," *Safety and Health*, National Safety Council, August 2002, p. 29.

3. Bureau of Labor Statistics, US Dept. of Labor, January 2002.

4. Bureau of Labor Statistics, US Dept. of Labor, January 2002.

5. Section 5(a)(1) of the OSHAct of 1970.

6. Ethanol 2000, LLP, Bingham Lake, Minnesota.

7. 29 CFR 1910.120(a)

8. Adapted from HAZWOPER Response Levels, Coastal Video Communications Corporation, Virginia Beach, VA, catalog number HWO02H.

9. "Process Technology," *Chemical Process Safety Report*, Thompson Publishing Group, 1725 K St. N.W., Suite 700, Washington, D.C. 20006, 1996.

10. Selections from "Process Technology," p. 301-311.

11. The following summaries are adapted from *Accident Prevention Manual for Business & Industry, Environmental Management, 2nd edition*, National Safety Council, p. 65-74.

12. www.epa.gov/compliance/civil/federal/caa.html

13. www.epa.gov/compliance/civil/programs/caa/caaenfstatreq.html

14. *Accident Prevention Manual for Business & Industry, Environmental Management,*, p. 72. Reprinted with permission.

15. Exceptions are currently under study for flammable liquids stored in atmospheric tanks or transferred while kept below their normal boiling point without benefit of chilling or refrigeration.

16. www.census.gov/naics

17. 40 CFR 68.50, Accidental Release Provisions, Subpart C.

18. "Security Study, An Analysis of the Terrorist Risk Associated with the Public Availability of Offsite Consequence Analysis Data under EPA's Risk Management Program Regulations," EPA 550-R97-003, US Environmental Protection Agency, Washington, DC, Dec. 1997.

19. Walter L. Frank, P.E., and Steve Arendt, "Reshaping Process Safety Regulations," *Chemical Engineering Progress*, March 2002.

20. Corporate Written Safety Plans, Envirowin® Software LLC, www.envirowin.com.

Safety and Health Training

Introduction

Plants make money through the products they produce, and to a large extent, the amount of profit depends on plant productivity and product quality. To maximize these factors, facilities often rely on "process-product training." Without a good process understanding and knowledge of how to produce consistent product features, the organization cannot survive the competition for very long.

Safety is an important strategy to gain competitive advantage. While facility leaders currently spend a lot of money on environmental, health, and safety (EHS) training, many find it difficult to determine what benefits they receive in the way of increased performance. The cost of EHS training is sometimes looked upon as a "tax," or as a cost of doing business rather than as another way to achieve a competitive advantage. After all, only a reliably safe process is capable of producing a quality product at low cost. An effective EHS training system is critical to gaining this competitive business position.

Safety training is too often done in a vacuum. In other words, safety training should not be separate from other training. One of the goals of this chapter is to demonstrate the importance of integrating process-product and chemical safety, health, and environmental training. This integration strategy is especially important in the chemical industry because daily hazardous material handling often becomes so routine that it replaces safe work practices.

Training Issues and Strategies

As a result of a rapidly changing workplace and other factors, there are many challenges to managers of industrial EHS training systems. Consider the following list of issues:

1. Safety training is often under-planned and over-done without assessing organizational and individual needs.

2. The training is used as punishment for an accident, near miss, unsafe act, or failure to follow procedures rather than to build employee knowledge and skills.
3. EHS training is often confined to required regulatory topics and done separately from higher priority product-process training.
4. It is difficult to quantify the return on investment in EHS training.
5. Despite recent efforts, there continues to be an undersupply of trained and experienced safety and health professionals as candidates for facility jobs.
6. Many employees cannot read well enough to understand EHS regulations or corresponding facility programs.
7. Many employees perceive themselves as more likely to be injured at home or at play than at work. Thus they do not take safety training seriously.

Integrate product-process and safety training around important facility safe work practices. For example, the primary product for the plant may be a flammable or toxic liquid. The assessment should show the need for employees to learn work practices that eliminate and reduce exposure to the liquid, its vapor, and ignition sources. Much of the training will then involve handling the product and intermediates in closed systems (vessels, columns, pipes, etc.). When open handling is required (e.g. during sampling procedures), the training needs to reinforce use of engineering controls, safe work procedures, and appropriate personal protective equipment (PPE).

Employee involvement in any program is necessary for success, and this is especially true for safety training. For this reason, the safety manager must take a cultural and behavioral approach (see Chapter 7). Use training, feedback, and positive reinforcement of safe work practices to achieve "habit strength." This approach will result in movement from 1) Regulatory compliance, to 2) Company policy and procedures, to 3) "I want to work safely."

Consider incentives for attendance or making attendance mandatory at first. Measure employee perception continuously; serious problems often arise when employees are forced to participate over a long period of time.

The Need for Chemical Process Safety in Education

One of the most powerful teaching tools is using actual case histories to reinforce basic safety principles. Safety and Chemical Engineering Education (SACHE) is a consortium of universities that receive teaching aids developed by the Undergraduate Education Committee (UEC) of AIChE's Center for Chemical Process Safety (CCPS).[1] The UEC has been conducting student essay contests since 1996 that promote blending process safety and loss prevention into undergraduate chemical engineering programs.

One prominent university has included a three-part safety training program into their curriculum. Classes include:

1. An introduction to safety and health through simple video-based case histories of fires, explosions, and toxic releases.
2. More technical lessons in the second and third years. Many involve evaluation of chemical process systems from a safety standpoint. Here students are introduced to thermodynamic, fluid mechanical, and mass transfer calculations that have an impact on safety.
3. A required senior year course in Process Safety Engineering that includes topics important to the chemical process industry such as avoiding fires, explosions, and toxic releases by providing proper ventilation, controlling static electricity, etc.

If more colleges and universities adopted similar programs, incidents would be reduced. There would be more candidates for chemical plant safety coordinator positions and a greater proportion of engineers would become certified safety professionals (CSPs) as they gained experience. This would accelerate the rate at which facilities could build and strengthen their chemical safety management infrastructure (see Chapter 11).

Management of a Chemical Safety Training System

Organizations need a management system to verify that employees understand chemical hazards, related process technology, and operating/safety equipment associated with their jobs. Sound adult learning principles should be employed that include assessing needed skills and knowledge, involving employees in setting learning goals, engaging activities directly related to on-the-job tasks, and relating the training to immediate needs.

EHS training programs should be designed to meet specific job functions in a plant. Functions should include chemical process operators, plant laboratory chemists and technicians, technical personnel, maintenance employees, supervisors and managers, and EHS personnel. Thus, it is critical to define key safe operating and maintenance procedures in order to determine learning outcomes.

Plant managers and supervisors need to assess employee job qualifications and establish development plans that include EHS objectives. Safety should be a part of the performance appraisal and development process (PADP) for all new and experienced employees.

Specification for and documentation of all training programs is needed. A feedback system must also be established for employees and plant management. The system should include an evaluation procedure to verify that the training program meets EHS objectives.

Renewal and reinforcement of the training program is also necessary. Periodic refresher training courses, which are current and appropriate to experienced employees, should be provided. It is equally important to provide a system that incorporates changes and new EHS information. This includes describing lessons learned from chemical process accidents and near misses into existing training programs. The rest of this chapter, then, is organized into an implementation process for installing a chemical safety management training system:

1. Assessment of training needs and planning
2. Select initial training program
3. Develop and field test program
4. Implement training and assess outcomes

Implementing a Training System

Step 1: Planning

A common problem throughout industry is purchasing a training package without first assessing organizational and individual needs. A key planning requirement is to link the training to all facility functional operating and maintenance procedures. Assessing needs and basing the training on relevant procedures must be part of goal and objective-setting processes.

Assessing Employees' Skills and Training Needs

Imagine the following scenario. A chemical operator violates the plant lockout-tagout rule in preparing a vessel for repair. He cleans and rinses the equipment but neglects to lock out the agitator drive and blank/tag raw material and product lines. Without checking the system and applying his own locks and tags, the maintenance mechanic enters the vessel. He becomes ill after 45 minutes and has to be helped to climb out of the tank. Without much thought, the operations and maintenance supervisors get together and have both men retake the lockout/tagout (LOTO) training they completed 6 months ago (the plant does annual lockout/tagout training).

In this case, training was used as punishment rather than as a means of providing the two employees additional knowledge and skills. Additional training, in this example, was not the need. A better solution would have been appropriate disciplinary action for both employees and the establishment of a short-term performance improvement plan.

The National Institute for Occupational Safety and Health (NIOSH) defines the basic learning model[2] as:

Training ⟶ Learning ⟶ Action

The goal is the action taken as a result of learning. That is, we want the employee to demonstrate new skills. Therefore, first document the skills that will be required to perform the jobs that have a direct impact on operational safety. Then employees can begin a well-defined development plan, and needed training programs can be determined. Skill assessments are best made before hiring or transfer from another job or during a probation period (either as part of initial employment policy or as a result of unsatisfactory or marginal job performance).

It is common in the chemical and other industries for employees to come in at the entry level and build knowledge and skills gradually through on-the-job and formal training programs. Through this process, individuals develop the qualifications necessary for jobs with increasing responsibility and the organization strengthens its operational and safety infrastructure. Define job skills needed over the next five to ten years for operators, maintenance personnel, contractors, technical personnel, emergency response people, and supervisors/management. Facilitate this process by doing the necessary planning early.

Table 9-1 Procedures Related to Chemical Operator Performance
Requirements

Performance Requirements	Example Procedures
1. Make process and equipment adjustments to safely control unit efficiency, productivity, and product quality within prescribed limits.	Follow steps to avoid deviation from upper and lower quality/safety limits using computer based distributed control system (DCS).
2. Inspect and observe equipment and process operation for abnormalities.	Follow procedure to respond to high tank level alarm.
3. Troubleshoot equipment and process operation when abnormalities are recognized and take appropriate corrective action.	Follow emergency shutdown procedure for rupture disk blowout in chemical reactor.
4. Immediately and correctly respond to any critical operating parameter (COP) reaching its "never exceed" limit.	Initiate emergency shutdown and chemical neutralization procedure for pressure near mechanical limits of equipment.
5. Recognize, analyze, and correctly respond to emergency situations.	Respond to chemical splash on coworker- sound alarm and initiate emergency procedures from MSDS.
6. Make minor repairs and conduct routine maintenance given the necessary authorizations.	Replace chemical hose (replacement-in-kind) feeding solvent to mix tank.
7. Identify items in need of maintenance and initiate requests for work order.	Complete daily equipment checklist procedure prior to start up and initiate work order to replace defective pump.
8. Prepare equipment for maintenance.	• Initiate production lockout-tagout procedure. • Drain and block/bleed all chemical lines prior to confined space entry.
9. Provide timely, complete, and accurate written and oral communications.	Provide objective relevant facts during accident investigation interview.

Link Operating and Maintenance Procedures with Learning Requirements

Linking procedures to training is the best way to integrate safety, health, and environmental training with plant operations. In Chapter 6, we learned that tools such as Job Safety Analysis (JSA)/Job Instruction Training (JIT) can be used to document unwritten procedures. Operational and maintenance procedures having the greatest impact on safety should be in place (or at least in draft form) before selecting training programs. It is through procedures such as chemical sampling and line breaking, labeling of containers, and confined space entry that important chemical safety principles are communicated to employees.

Operating Procedures

Much of the success of the facility safety program depends on chemical operators and their safe work practices. Consider the example performance requirements and procedures in Table 9-1.

This critical procedure information is used to specify relevant training programs that will deliver the necessary knowledge and skills to function safely. From the table, the reader can easily see how important it is that training is customized to the procedural needs of the process and the plant. Too often, a generic set of slides, a video, or a multi-media program is pulled off the shelf and delivered to operators without change. The sure way to make training effective is to integrate safety and health with what your people do on the job every day.

Maintenance Procedures

Critical tasks for maintenance personnel are often related to assuring the mechanical integrity of equipment. Critical equipment from a safety standpoint includes:

- Tanks, columns, and especially pressure vessels
- Chemical process lines
- Vents, pressure relief systems, and other systems and components such as rupture disks and pressure relief valves, explosion relief walls, vacuum breakers, safety interlocks, as well as chemical detection and suppression systems

Table 9-2 details procedures and relevant training for mechanical integrity inspectors.

Table 9-2 Procedures and Training Required for Mechanical Integrity Inspectors

Inspector Level	Typical Tests and Procedures	Inspector Background	Training Required	Qualifications
1	External and internal inspections of reactors, tanks, condensers, rupture disks, relief lines, valves, pumps, piping, chemical detection and deluge systems. Refers defective items to level 4.	Chemical operator or craftsperson	Company and/or plant certification	Documented on-the-job demonstration of learned skills
2	Metal thickness, measurements, in-house calibrations, drive vibration and alignment checking. Refers defective items to level 4.	Operator, craftsperson, * or contractor	Company or UT certification plus level 1 training	Documented written test plus field demonstration of learned skills
3	Hydrostatic, wet fluorescent magnetic particle, radiography and eddy current tests; bench testing of pressure relief valves plus others as needed. Refers defective items to level 4.	Specialized contractor	Appropriate training in areas of specialty: API 510, 653, 570, etc.	Documented certification
4	• Responsible for assuring that all inspections and tests are complete and that documentation is up-to-date. • Coordinates and interprets all instrument measurements and equipment defects found. • Estimates remaining service life for equipment. • Confers with resident engr., plant engr., project engr., corporate pressure vessel coordinator and/or qualified contractor regarding repair, replacement, re-rating, or decommissioning of equipment.	Plant mechanical integrity coordinator	Levels 1, 2, 3 training plus training in interpreting instrument readings and calculating service life.	Documented written test, plus field demonstration of learned skills.
Resident engr., Plant engr., Project engr.	• Confers with plant mechanical integrity coordinator regarding instrument measurements and equipment defects found. • Confers with plant mechanical integrity coordinator regarding remaining service life. • Confers with corporate pressure vessel coordinator, contractor and plant mechanical integrity coordinator regarding repair, replacement, re-rating or decommissioning of equipment. • Functions as a resource person.	Degreed engineer	Levels 1, 2, 3 training plus training in interpreting instrument readings and calculating service life.	Documented written test, plus field demonstration of learned skills.

* ASME requires that operators and craftspersons qualified as inspectors do not perform inspections on equipment in the specific areas in which they work

Note that specialized external contractors often perform typical level 2 and 3 work. It is more cost-effective, in the long run, to train and certify employees to do as many of these inspections and tests as practical. Many plants will hire an inspection service as their process safety management program is first being established and replace external inspectors with employees as they build experience and skills.

Set Program Goals and Objectives

Program goals are general statements such as "Conform to hazard communication, process safety management, and other federal, state, and local requirements applicable to our facility." Another example is "Our goal is to outline policies and procedures that meet or exceed all regulatory requirements." Objectives are more specific and need to be achievable, observable, and measurable. Goal and objective setting is also an opportunity to describe the connection between safety and quality. Also remember to define EHS and other resources needed to accomplish the objectives.

Choose Topics, Training Medium, Level of Complexity, and Frequency

Many organizations make the mistake of segmenting their operational and safety training. They may devote many resources to topics such as specific unit and equipment operation, plant economics, and instrumentation/control systems. Basic environmental, health, and safety classes are taught separately and often confined to required regulatory topics using generic content.

It is a good idea to start with required topics and frequency (Table 9-3) to assure regulatory compliance.

Note that different departments (operations, commodities, maintenance, etc.) have different initial and refresher requirements, depending on the topic. This is a simple way to relate training to job duties.

Most often, OSHA requires training whenever an employee's job assignment changes. From a practical standpoint, most organizations conduct regulatory training annually and use OJT at the time of the job change. This is a good time for supervisors to involve experienced employees in training the new person.

Table 9-3 Safety Training Matrix[3]

	Process Safety Management Overview	Process Safety OJT	Hazard Communication	Fire Safety	Fire Safety Drill	Lockout – Tagout	Lockout-Tagout OJT	Confined Space Entry	Confined Space Entry OJT	Notes
Operations	initial / annual	initial / **3 yrs**	initial / annual	initial / **annual**	initial / **annual**	initial / annual	initial / 3 yrs	initial / annual	initial / 3 yrs	
Commodities	initial	initial	initial / annual	initial / annual	initial / annual	initial / annual	initial / 3 yrs	initial / annual	initial / 3 yrs	
Maintenance	initial / annual	initial	initial / annual	initial / annual	initial / annual	initial / annual	initial / annual	initial / annual	initial / annual	Program tbd.
Carbon Dioxide	initial / annual	initial / 3 yrs	initial / annual	initial / annual	initial / annual	initial / annual	initial / 3 yrs	initial / annual	initial / 3 yrs	
Boiler & Water	initial / annual	initial / 3 yrs	initial / annual	initial / annual	initial / annual	initial / annual	initial / 3 yrs	initial / annual	initial / 3 yrs	
Laboratory	initial / annual	initial / 3 yrs	initial / annual	initial / annual	initial / annual	initial / annual		initial / annual	initial	
Janitorial	initial / annual	initial	initial / annual	initial	initial	initial / annual	initial	initial / annual	initial	
Notes		Maint. - mch'l integr.	annual		annual					

Table 9-3 Safety Training Matrix (continued)

	Process Safety Management Overview	Process Safety OJT	Hazard Communication	Fire Safety	Fire Safety Drill	Lockout – Tagout	Lockout-Tagout OJT	Confined Space Entry	Confined Space Entry OJT	Notes
Office	initial		initial	initial	initial	initial		initial		Office ergonomics, etc.
	annual			annual	3 yrs					to be determined
Technical	initial	initial	initial	initial	initial	initial	initial	initial	initial	
	annual	3 yrs	annual	annual	3 yrs	annual		annual		
Management	initial		initial	initial	initial	initial	initial	initial	initial	
					3 yrs					
E Squad	initial	initial	initial	initial	initial	initial	initial	initial	initial	E Squad tbd
	annual	annual	annual	annual	annual	annual	annual	annual	annual	
Notes	MM	tbd by tech mgr	MM Cust.	MM Cust.	tbd	MM Cust.	tbd	MM Cust.	tbd	

Bold   training frequency defined by regulation
MM: Multimedia
Cust.:   Customized

Integrating Operational and Safety Training

A common problem in safety training is failure to customize generic pro-grams such as lockout-tagout of hazardous energy sources to meet the needs of the facility. Clear procedures and training for chemical line-breaking are also essential parts of the organization's lockout-tagout policy.

The requirements determined by OSHA's PSM regulation are very help-ful to integrate operational and safety training. All 14 elements of PSM, including training, need to be blended into the plant's process safety man-agement system. For example, consider the requirement for training on operating procedures. Most chemical plants' operating controls are part of a computer-based distributed control system (DCS). Therefore, site-specific training for operating procedures comes not from the purchased compact disk (generic content), but from in-plant sit-down situations and on-the-job training using the plant DCS operations manual. So while a multi-media safety presentation on CD-ROM provides an overview of what is required, the customized details must come from within. This example stresses the need to integrate operational and safety training, especially for facilities handling chemicals (Table 9-4).

Operators and maintenance personnel, technical and safety personnel, and management each have training programs designed to job perfor-mance requirements like those for chemical operators found in Table 9-1. Note that the procedures and performance requirements are chemical, process, and equipment-specific. This means it is critical that the train-ing be site-specific and nearly independent of topic. For example, an off-the-shelf process safety management program has little value if it is not integrated with operational training for plant processing of anhydrous ammonia. Safe work practices such as management of change or line-breaking procedures will be different for ammonia than an ethanol or sulfuric acid plant.

Consider another example. The author was involved in helping a client integrate PSM training with a 100-question operations quiz. The quiz is designed to improve process and equipment understanding for new and experienced operators. Safety and health considerations were folded into the quiz by adding questions for covered process sections relative to

*   Design intent and process chemistry
*   Critical quality, safety, and operational parameter limits for tanks/vessels, columns, lines/pipes and pumps, and heat exchangers
*   Safety and health considerations from material safety data sheets

Table 9-4 Integrating Operational and Safety Training

Topic	Operators, Maintenance[2]	Technical, Safety[3]	Management[1]
PSM Management Practices[1]			
Empl. Participation	X	X	X
Training	X	X	X
Contractor Safety		X	X
Audits	X	X	X
Trade Secrets	X		X
Engineering Controls[1]			
Process Safety Information	X	X	
Process Hazards Analysis	X	X	
Safe Work Practices[1]			
Management of Change	X	X	X
Mechanical Integrity	X	X	X
Pre-start Review	X	X	X
Operating Procedures	X	X	
Hot Work	X	X	
Lockout-Tagout	x	X	
Line Breaking	X	X	
Confined Space Entry	X	X	
Incident Procedures[1]			
Incident Investigation	X	X	X
E. Planning & Response	X	X	
EHS Operations, Procedures	X	X	X
Chemical Reactions, Reactors	X	X	
Plant Economics	X	X	X
Envir. Sound Operations	X	X	
Unit Operations	X	X	
Equipment & Problem Solving	X	X	
Utilities & Services	X	X	
Instr. & Control Systems	X	X	
Project & Program Mgmt.		X	X
Interpersonal Skills		X	X
Performance Appraisals		X	X
P. Computer Applications		X	X
Pr. Solving, Decision Making		X	X
Productivity Management		X	X
Stress Management	X	X	X

[1]Adapted from Process Safety, product code STPSMMPG (discontinued), Mastery Technologies, Inc., Farmington Hills, Michigan and Richardson Texas, www.masterytech.com
[2]Adapted from Training Planner for Chemical Industry, Synthetic Organic Chemical Manufacturers Association (SOCMA), Washington, DC, www.socma.com.
[3]Adapted from Appendix 10C, Plant Guidelines for Technical Management of Chemical Process Safety, Center for Chemical Process Safety (CCPS), American Institute of Chemical Engineers, New York, N.Y., p. 221.

The integrated training uses a PowerPoint format with audio capability plus provisions for video, diagrams, and still photos. The program is delivered as self-study, on a one-on-one basis, or in small groups. The distributed control system can be used to demonstrate important procedures such as response to alarms, normal and emergency shutdown, and response to loss of power.

There are countless industrial packages on the market that claim to meet all the environmental, health, and safety training needs for the plant. But remember that the most credible evidence of training comes from the inside rather than the outside. Whether employees are hired based on their education and training and/or receive training on-site, they prove that they have the necessary qualifications for a job by doing it. Thus, the temptation to buy a so-called "off-the-shelf" EHS training package must be resisted unless the intent is to customize the program to the facility as well.

Furthermore, real learning comes from a variety of sources, and formal training is only one of them. Supervisors and training coordinators normally put a family of learning experiences together to help employees meet job qualifications and to keep them growing into greater areas of responsibility.

When it comes to safety training, the situation is the same. Employees learn to work safely not only through completing course work, but through drills, participating on-the-job with an experienced person, and by other means. Thus when choosing EHS training materials, we must remember to look ahead to how the materials will be integrated with other learning experiences to meet job qualifications and regulatory requirements.

Selecting Training Medium

Selecting the right training medium is dependent on important considerations such as the level of complexity, degree of customization, flexibility in scheduling, provisions for feedback, and testing, as well as required development resources (Table 9-5).

Another key consideration in selecting medium is matching the content and essential data to be communicated to expected outcomes from the learner. Consider the following four levels of thinking:

1. Information—learned by drill and practice
2. Judgment
3. Interpretation
4. Evaluation

Table 9-5 Comparison of Conventional and Multi-Media Training Systems

	Instructor-Led	Videos & Films	Multi-media CBT
Capabilities	• Customization • Large volume of information • Open-ended learning • Strong group process • Ideal for role playing	• Ideal for factual and visual information • Ideal for event simulation • Good for persuasion- can trigger emotions	• Wide range-application • Individualization and interaction • Branching and self-paced • Provides feedback and testing • Provides task-based instruction and practice • May be customized
Advantages	• Flexibility of content, methods and techniques • Social benefits • Rapid development • Economy if internal	• Good for specific how-to knowledge • Reinforces people and objects in motion • High visual impact • Can include many types of materials	• Versatility/efficiency • Economy/convenience • Consistency • Reinforces correct learner response • High learner retention • Ease of record keeping
Limitations	• Inconsistency • Limited feedback • Pace may be controlled by slow learners • Scheduling limitations	• One way communication • Not paced to the learner	• Decision makers may have unrealistic expectations • Simulation; not hands-on • Limited content • High development costs • Long development time • No human interaction • Inappropriate for groups

"Information requires the lowest level of thinking. Judgment, interpretation, and evaluation involve more complex critical thinking and problem solving skills."[4] For example, emergency response to a chemical spill is effectively taught via an instructor-led classroom session (20 minutes–1 hour, depending on the audience), followed by a hands-on drill (2–4 hours). Process hazards analysis training usually requires several days using several mediums to teach higher skill levels.

Conventional Learning Systems

Most readers will relate to the following examples of conventional EHS learning systems

- On-the-job equipment and safe work practice training for chemical operators
- Drills and exercises such as the use of fire extinguishers, practicing evacuation, and shelter-in-place procedures, etc.
- Instructor-led classroom training such as discussion of hazard communication, chemical exposure limits, and use of PPE
- Use of videos such as HAZWOPER or Confined Space Entry for training in group settings
- Case history workshops

All experienced EHS trainers have battled the scheduling problems associated with instructor-led training. Assume, for example, the topic is hazard communication and the trainer is an industrial hygienist that lives two hours away. Your plant has 50 employees and you plan to schedule half of the participants on a Friday afternoon and the other half two weeks later. But only 18 of the first 25 scheduled attend the training, due to vacations, illness, and personal or family conflicts. It might take two or three alternate dates to get the remaining seven people through the course. If similar problems occur with the second scheduled session, the time lost and additional expenses may exceed the training budget due to the cost of four to six additional sessions. Finally, this hassle may be for only one of six or seven topics on the refresher schedule for the year.

Instructor-led training, video and film training, and on-the-job training have been around a long time. These conventional systems have many capabilities and advantages—including low cost (especially if internally developed and delivered) and flexibility. NIOSH calls slides and overheads prepared by the instructor, "low-risk curriculum enhancements" because they are simple and inexpensive to produce.[5] Availability of digital cameras and internal CD burners has significantly improved the quality of do-it-yourself training.

Videos and films are ideal for factual and visual information, e.g. depicting case histories of major fires. Many types of materials can be included in a video, and videos are relatively low cost and long-lasting. However, videos are not interactive and the pace is beyond the learner's control. Some employees used to being on their feet and doing things with their hands often do not do well sitting in a darkened classroom, watching a video they have seen several times before.

Using Case Studies as Training Materials

Short of direct involvement in an accident, there is no better teaching tool than a case study. Participants generally cannot help identifying with the people involved, and if relevant background and resource information is provided, a case study becomes a powerful "lesson learned."

The Center for Chemical Process Safety (CCPS) is involved in a project to provide case study packages to university classrooms. Case studies can be used for industrial training seminars if the sponsoring group joins SACHE (Safety and Chemical Engineering Education, mentioned at the beginning of this chapter). Roughly 100 universities belong to SACHE and pay an annual fee of $300. CCPS funds subsidize the development of teaching aids.

Table 9-6 provides the current SACHE listing.

The case study packages available through CCPS consist of slides and easy-to-use lecture notes. Videotapes are included when available.

The presentation on fires (item 2, Table 9-6) includes a flammability diagram for methane. This is a good tool for demonstrating to new employees and technicians how information from a Material Safety Data Sheet can be used for fire prevention.

Other CCPS case studies include:

Seminar on Tank Failures
- How steel can be embrittled at low storage temperatures for liquid natural gas
- Failure of old, defective welds due to filling a tank on an extremely cold day
- Tank failure due to siphon action

Table 9-6 SACHE Case Studies Available from AIChE-CCPS[6]

Title	Author	Year	ISBN Number
Seminar on Tank Failures- slides and lecture	Willey	1993	ISBN 0-8169-0602-5
Fires- slides and lecture	Welker/ Springer	1993	ISBN 0-8169-0603-3
Explosion Proof Electrics- slides and lecture	Cyanamid/ Page	1993	ISBN 0-8169-0599-1
Process Safety Management With Case Studies: Flixborough And Pasadena (TX) And Other Incidents- slides and lecture	Bethea	1994	ISBN 0-8169-0608-4
Seminar On Nitroaniline Reactor Rupture- slide and lecture	Willey	1994	ISBN 0-8169-0634-3
Seminar On Seveso Release Accident Case History – slides and lecture	Willey	1994	ISBN 0-8169-0634-4
Dust Explosion Control – video/slide/lecture	Louvar/ Schoeff	1994	ISBN 0-8169-0634-4
Toxicology And The Chemical Engineer- slides and lecture	Welker/ Springer	1995	ISBN 0-8169-0606-8
Consequences Of Operating Decisions- lecture	Cobb	1995	ISBN 0-8169-0633-5
Industrial Hygiene And The Chemical Engineer- slides and lecture	Springer/ Welker	1995	ISBN 0-8169-0604-1
Phillips' Explosion- video	Bethea	1996	ISBN 0-8169-0673-4
Inherently Safer Plants- slides and lecture	Kubias	1996	ISBN 0-8169-0669-6
Property Of Materials – slides and lecture	Willey	1997	ISBN 0-8169-0694-5
The Bhopal Disaster- video/slide/lecture	Willey	1998	ISBN 0-8169-0766-8
Potential Accidents From Safety Systems- slide and lecture	Hendershot	1998	ISBN 0-8169-0732-3
Emergency Response Planning- slide and lecture	Bethea	1998	ISBN 0-8169-0671-8
The Human Health Risk Assessment Process- slide and lecture	Jayjock	1998	ISBN 0-8169-0734-X

The Bhopal, India Disaster
- Entry of water into a methyl isocyanate (MIC) system causing exothermic reactions
- Importance of functional backup safety systems such as vent gas scrubbers and flare stacks
- Use of a slip blind to isolate water from the MIC

The Seveso, Italy Release
- Extreme toxicity of materials like dioxin
- Mechanisms of self-heating and runaway reactions
- Consequences of operating decisions deviating from written procedures
- Dispersion patterns and community impact from release of dioxin

Nitroaniline Reactor Rupture
- Importance of a management of change system
- Need for redundant safety component (pressure gauge between relief valve and rupture disk)

Phillips 66 Explosion
- Effectiveness of a facility emergency response plan
- How information about an accident evolves and becomes reported by the media

Advanced Learning Systems

In today's world, the current wave of advanced learning is computer-based, distance learning. There are many variations of computer-based training (CBT) that provide attractive features like those in Table 9-5:

- Just-in-time and self-scheduling based on need
- Multi-media features—video, sound, and animation
- Built-in testing
- Bookmarks indicating where the user stopped prior to interruptions[7]
- Ability to customize the content to address facility policies and procedures

Primary advantages of multi-media computer-based training (MMCBT) are absolute consistency, integrated testing and recordkeeping, and provisions for self-scheduling. In one plant that implemented a multi-media computer-based training system, EHS and training coordinators established one safety topic per month for each process building. Employees then had 30 days to schedule their own individual session. Process team leaders were responsible to see that each employee completed the training needed as required by their performance appraisals and development plans. Duplicate training sessions were eliminated.

There are important limitations to computer-based training (CBT) as shown in Table 9-5. Sometimes these systems are oversold, with claims that CBT is "all things to all people." Or decision-makers reach unrealistic expectations about CBT. A CBT program can be complex and often requires a high development cost. For this reason, NIOSH considers video and CBT "high-risk curriculum enhancements."[8]

Another limitation of CBT is the absence of human interaction. Someone is needed to debrief the employee or answer his or her questions not addressed adequately in the CBT interaction. However, as employers do a better job of integrating CBT into the plant training system, solutions to this human interaction issue are coming into use.

In a 1999 decision, OSHA responded to an inquiry by saying, "During training, it is critical that trainees have an opportunity to ask and receive answers to questions where material is unfamiliar to them. Frequently, a trainee may be unable to go further with the training or to understand related training content until a response is received."[9] The agency goes on to state the requirement to have "direct access to a qualified trainer" during training, for example, via a "telephone hotline." OSHA does not consider use of an e-mail system to be direct access. In this author's opinion, CBT is so important to current and potential users that the direct access issue can and will be addressed to satisfy OSHA.

We commented earlier, that it is important to resist the temptation to buy an "off-the-shelf" training package and expect it to solve all training issues. The key to a successful approach is that the training medium and content is based on a thorough organizational and individual needs analysis.

The final components of planning are evaluating choices for measurement and documentation systems.

Measurements

One of the issues identified early in the chapter is ability to quantify safety training outcomes. It may take years for training along with other elements of a strong chemical safety management infrastructure to impact the recordable and lost-time incidence rate. Look for injury and work-related benefits in the long term. Put performance indicators in place (Chapter 7) including things such as topics completed and training as scheduled that will help measure continuous improvement. For chemical safety, consider outcome measures such as reduction in spills, releases, and fires; fewer emergency shutdowns; and process incidents/upsets resulting from operator error.

Documentation Systems

OSHA generally requires that documentation include the identity of the employee, date of training, and means used to verify that employees understood the training. However, the organization will benefit by designing a centralized documentation system even if it is not fully implemented immediately. For exempt employees, it is a good idea to design the system to include safety and other training as part of the employee personnel development file. With increasing electronic capability, it is easy to document an employee needs analysis, course outlines and trainers, course feedback, revision and refresher information, and measures of pre and post-training performance.

Step 2: Select Initial Training Program

During this step, focus on essential content, data, and information to be included in the course. Base the learning objectives and content on facility operating procedures and regulatory requirements. Determine the level of complexity and choose training medium(s) that will accomplish expected outcomes. Draft instructional materials using low-cost visual aids such as simple video camera work, overhead transparencies, and slides. Consult subject matter experts to help define training content and identify preferred mediums. Interview managers and training consumers to assure that the program meets identified needs.

Training Specifications

It is important to have a training specification (at a minimum, draft key work procedures) before choosing a training program for the instructor to deliver. The program learning objectives need to reflect safety and health considerations from the job/task analysis (i.e., JSA and JIT). The specification then integrates the specifics into a single document.

Develop a training specification for internal use and communication to external providers. The specification should include course title and code, participating departments, course frequency, schedule and length, levels of complexity, learning objectives, criteria for course completion, and regulations covering the topic.[10]

Step 3: Develop and Field Test Program

After all the above work is completed, it is tempting to simply jump ahead and start the new training program. But invariably, changes are needed that are difficult or expensive to make if development and field testing is bypassed.

The key factor here is to conduct a field trial before producing complex and high-cost enhancements. Make one or two revisions using overheads or handouts before producing a high-quality video or specifying a customized, multi-media delivery system. Carefully document the results of each improvement using the learning objectives and assessment criteria. Make a timeline (objective versus date due) for producing the final training program.

Step 4: Implement Training and Assess Outcomes

At last, the curtain rises and the training program begins. Now, the effectiveness of all the planning, development, and production is apparent.

One example that demonstrates the significant effect good planning and testing can have on the implementation of effective training is emergency squad training conducted at a 3M "smoke house." Here, the building is a completely darkened double-garage containing partitions, 55 gallon drums, and other obstacles, sealed well-enough to hold a moderate level of "artificial smoke." OSHA's regulations for employee rescue (29 CFR 1910.38), fire brigade training (1910.156), and respiratory protection (1910.134) were used as guidelines. Self-contained, air-supplied (SCBA) respirators were used to protect the rescuer-trainee and facilitate movement in the smoke-filled facility. Trainees (5–10 participants) were to enter and exit the smoke house in teams of two, carefully following oral instructions on how to systematically work their way through the building. Certified safety professionals were placed inside as observers.

The qualitative respirator fit test required detailed preparation of an isoamyl acetate (IAA, banana oil) and odor-free water solution. There were further instructions for evaluating the employee's comfort while wearing the respirator (5 minutes) and use of the banana oil to test the seal of the face mask. Basically, if the employee could not detect the banana oil odor, he or she passed the fit test.

The point of this story is that without considerable behind-the-scene preparation, this hands-on training exercise would have been a miserable failure. In order to safely do the fit testing and conduct the smoke house exercise in a timely manner, a written procedure was developed for preparing and using the IAA. Further planning and field testing helped streamline the process of filling the building with artificial smoke, donning and testing the SCBA units, and placing the observers.

Training program evaluation is important to objectively measure that the course contributes to safe operations, that the course is cost effective, and that a feedback mechanism is in place to continuously improve the program.

The assessment procedure must include measuring the performance of the students, the course itself, and the instructor. Does the training result in increased knowledge, and how is that demonstrated? For example, the objectives of a process safety management training program for maintenance employees are:

- "To provide a safety net around the performance of critical work
- "To establish location-specific safety, environmental, and operational requirements for the performance of critical work—that is, work that has the potential for directly impacting the operational safety of the unit/location and for which no other safety net exists
- "To establish an effective and documented training program that has the confidence of management and satisfies corporate and regulatory requirements"[11]

Assessing training outcomes according to the criteria above results in improved training productivity.

Establishing an Instructor Program

Training programs do not run themselves, whether they are a conventional or advanced learning system. Because we are measuring training outcomes instead of "course completions," selection of the right instructor and/or coordinator-resource person can make or break the success of a program. This is especially important for a small facility where employees may not have either formal training or a safety department.

Instructor-coordinators may come from external sources or may be a full or part-time employee that has the right qualifications. Larger facilities frequently have some sort of "designer-trainer" function that helps to preserve and advance process technology. For example, the designer-trainer may be a chemical operator with extensive experience running multiple process units and is given a short-term assignment (six months to two years) to upgrade operations manuals used for on-the-job training (OJT). Or the facility could designate a maintenance training coordinator that has reached the fourth mechanical integrity inspector level, depicted in Table 9-2.

In choosing an internal instructor-coordinator, smaller plants cannot be as selective as larger facilities. However, management can still make a good choice if they pick experienced personnel who can perform a good job/task analysis and describe safe work procedures in simple terms.

Conclusion

Training is never the total answer to every performance gap and that is equally true in the business of chemical safety management. So-called traditional safety programs have suffered from a history of trying to shape safe work practices by "enforcing the rules." Even now, there are organizations that pay little attention to quality of instruction or follow-up to gauge training effectiveness. Training outcomes are difficult to measure directly. In addition, employees are more often seeking entertainment than motivated by the right to know (hazard communication). Employees are often given safety and health responsibilities without adequate qualifications. Site safety professionals are frequently so burdened by training demands that they cannot give adequate attention to strengthening the EHS infrastructure in other ways.

However, safety, health, and environmental practices are improving. Application of computer based learning technology is growing to make distance education a reality. Sources include computer-based training (CBT) via the internet, company intranets, or multi-media systems. Work done by The National Institute for Occupational Safety & Health (NIOSH) is leading OSHA to accept student "access to a qualified trainer" when not physically present. The agency's on-line small business outreach training program is providing more useful information about required topics. CBT, particularly multi-media systems, is often more effective than conventional delivery methods because it engages the learner, requires interaction, and offers feedback plus testing.

Advocating use of behavioral methods, work psychologists are helping organizations to eliminate negative attitudes about training and safety in general. Workers have more opportunity to contribute and participate in developing course content. Greater use of OJT, integrating operational and safety training, and the emergence of highly motivated self-directed work teams have improved safe work conditions and practices.

Effective training takes good planning and execution. Education and training must be systematic. An employee learning and development system is an essential element of chemical safety management (Figure 10-1). The line organization is responsible for training outcomes, even when the course is delivered by someone else.

There are substantial benefits from effective safety training. Increased productivity and efficiency plus greater quality result. Employees learn safe work practices that prevent injuries, work-related illnesses, and potentially catastrophic incidents. Good training improves morale and demonstrates the organization's commitment to safety, health, and environmental stewardship.

References

1. Lucas Osborn, "A SACHE Essay: Process Safety in Education," *Process Safety Progress* (Vol. 18, No. 4), Winter 1999.

2. *A Model for Research on Training Effectiveness*, U.S. Department of Health and Human Services, Public Health Service, Centers for Disease Control and Prevention, National Institute for Occupational Safety and Health, p. 7.

3. Ethanol 2000, Bingham Lake, Minnesota. Reprinted with permission.

4. Marcella Thompson, "Planning, Writing & Producing Employee Education Programs," *Professional Safety*, American Society of Safety Engineers, December 2000, p. 34-35.

5. *A Model for Research on Training Effectiveness*, U.S. Department of Health and Human Services, Public Health Service, Centers for Disease Control and Prevention, National Institute for Occupational Safety and Health, p. 11.

6. Ronald J. Willey, "SACHE Case Histories and Training Modules," *Process Safety Progress* (Vol.18, No. 4), Winter 1999. Reproduced with permission of the American Institute of Chemical Engineers. Copyright (1999) AIChE. All rights reserved.

7. R. Scott Lawson, "Computer-Based Training: Is It the Next Wave?" *Professional Safety*, June 1999.

8. *A Model for Research on Training Effectiveness*, p. 12.

9. "OSHA Responds to Letter Defining Online Training Requirements," www.coastal.com/hottopic.htm, Oct. 1999.

10. Thompson, Table 4, New Employee Safety Orientation Program Plan.

11. *Plant Guidelines for Technical Management of Chemical Process Safety*, Center for Chemical Process Safety, American Institute of Chemical Engineers, p. 213.

Implementing a Chemical Safety Management System

Introduction

A facility may have the best-written program in the world but it means little if it is not put into practice. It is essential to remember that safety is as manageable as anything you do in the plant. In this chapter, the reader will continue the three-step approach to improving safety performance.

Regulatory Compliance $\longrightarrow$ Company Policy $\longrightarrow$ "Want to"

Moving from regulatory compliance to policy is the *hard side* of safety management. This process turns regulatory compliance into conformance to company policy. More importantly, implementation of company policy builds a solid safety infrastructure within the organization. Change from policy to habit or "want to" represents progress on the *soft or people side* of safety management. A complete chemical safety management system implements a combination of hard and soft side program elements.

Consider a simplified example. The plant receives an OSHA citation for having no written personal protective equipment (PPE) program. So the initial emphasis is placed on meeting the OSHA PPE standard. That means that a hazard assessment must be done in order to specify the proper PPE to prevent exposure. Step two above involves implementing job safety analysis (JSA) as plant policy to document hazardous or infrequent procedures and tasks. The PPE required comes from the hazard assessment (Figure 8-3). Step three involves the cultural or people approach. Employees need a "want to" attitude about wearing PPE and working safely. Using data from observations, management can provide feedback and reinforce behavior to improve use of PPE. The message to employees, visitors, and newcomers becomes "This is the way we work around here."

Note that the infrastructure needs to be in place before taking the behavior-based systems approach. When do you know you are ready for a behavior-based system? Be driven by accident and investigation information.

Your basic Environmental, Health, and Safety (EHS) infrastructure is in place if people are surprised by accidents. Meanwhile, develop policy and keep chipping away at the hazards and issues.

There is a great return on investment from improving safety. In one recent study, 61% of the executives asked, believe their companies save $3 or more for each $1 invested in workplace safety.[1] The survey goes on to show that the investment goes beyond saving dollars. Seventy percent of the executives reported that "protecting employees is a leading benefit of workplace safety." Thus, implementing and continuously improving effective safety systems has hard and soft side benefits.

Major Elements of Chemical Safety Management System

There are many factors to consider in developing an infrastructure for chemical safety management. For example, what is the nature of your company or organization management structure? Do you operate a federal facility? Is this a gas or electric utility organization? What are our central/corporate and facility cultures like? What are applicable regulatory standards and what do our existing safety, health, and environmental policies and procedures look like? How far along have we come on the hazard identification, evaluation, and control learning curve?

Organizing the answers to these and other questions into nine elements of chemical safety management, we have a system that looks like Figure 10-1.

Design goals, objectives, and measures of outcomes and actions around each of the elements. Implement the elements in a parallel fashion, keeping the approximate order indicated. Focus especially on programs that control facility hazards and risks. Once the EHS structure is substantially in place, begin working to shape safe work practices into *habit-strength* behaviors.

Given these qualifications, facility management needs to take nine steps to fully install a chemical safety management system. They are:

1. Establish values, a vision, commitment, and annual improvement plan for safety.
2. Appoint an EHS coordinator(s) and provide adequate staffing and resources.

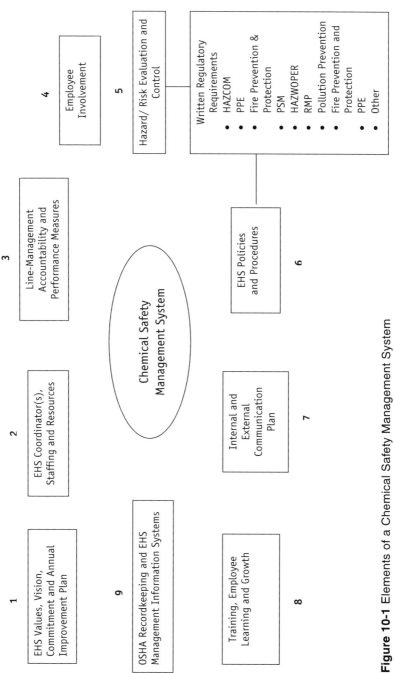

Figure 10-1 Elements of a Chemical Safety Management System

3. Define line-management accountability for safety. Provide performance measures and indicators.

4. Initiate opportunities for employee involvement.

5. Develop a system of hazard-risk evaluation and control measures.

6. Draft EHS policies and procedures that meet or exceed regulatory requirements.

7. Initiate internal and external communication plans.

8. Provide training and opportunities for employee learning and growth.

9. Strengthen OSHA recordkeeping and develop a functional EHS management information system.

Do *not* think of implementing the above steps in strictly a *sequential* manner. It is good to generally approach this project in the order indicated, but it could take years to finish if one step is not started until the preceding one is completed. Instead, begin as many of these steps as practical in a *parallel* fashion, recognizing that this is more about continuous improvement than beginning and finishing short-term tasks in sequence.

Step 1: Establish values, a vision, commitment, and an annual improvement plan for safety.

To take the first step in driving change, write down top management's values about safety, health, and environmental stewardship that apply to the facility. Consider values such as:[2]

1. Safety is a critical part of our business plan and will be considered in making all business decisions. Safety is as important as productivity and the quality of the products we produce. Only a safe work system produces quality products in a productive, cost-effective manner.

2. We will meet or exceed all applicable regulatory codes and standards.

3. All accidents are preventable. We will seek to eliminate all lost time; recordable, first-aid injuries; and work related illnesses. We will work to eliminate all incidents without injury, near-accidents, and potential hazards.

4. Line management has direct responsibility for the safety and health of the workers and other resources assigned to them on a day-to-day basis. Safety cannot be delegated to the safety coordinator.

5. Safety is a condition of employment. All employees are required to follow facility safety policies, procedures, and safe work practices.

6. All employees, supervisors, and managers will receive safety training that applies directly to their jobs to the fullest extent possible. Writ-

ten job descriptions and on-the-job training will include responsibilities for safety.

Once the values are on paper, the top facility manager will have an easier time drafting a safety vision that others can buy-into. Figure 10-2 illustrates the relationship between the vision and four important functional areas for safety: 1) culture and systems; 2) behavior; 3) safety programs; and 4) learning and growth.

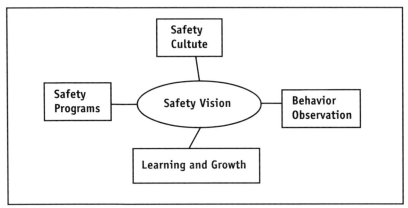

Figure 10-2 Safety Vision[3]

Note that these four functional elements drive the entire continuous improvement process.

Divisional and staff management understanding and support are also important to developing a vision and a set of goals and objectives for chemical safety management. Ask senior management to participate in developing a vision statement to the extent they are willing.

A good approach to drafting a vision is to get a group of senior people together. Ask them to push their chairs back, close their eyes, and ask individually "Given our values, what would this place look like a few years from now, if we had achieved our goal to improve safety?" Then individual ideas are recorded, discussed, combined, and finally summarized into a draft vision statement.

You will want to involve employees later, seeking buy-in to the vision statement as you roll out the internal communication plan. Without buy-in or at least acceptance of the vision, employees will have no real ownership of the safety improvement process.

Your vision statement might read something like "Safety here is first among the equals of safety, productivity, and quality. We expect our people to work safely, teach and encourage others to work safely, and to work together to build safe work conditions and practices."

Culture and Systems

Culture is basically the set of unwritten rules that determine "how things are done around here." Culture is clearly the most powerful factor that determines the degree to which an organization can achieve safety and health success. Although it may take a while to change significantly, facility culture is strongly influenced by the values of its leadership. Thus management must structure the chemical safety management system and shape the culture to be compatible. Integrating culture and management systems is an essential implementation strategy.

Behavior and Observation

An unsafe act like disabling a high level alarm on a mixing tank is an example of at-risk behavior. Management needs to identify unsafe/safe behaviors (Chapter 6) and use job safety analysis or behavioral methods to achieve improvement.

It is important for employees to go beyond regulatory and policy compliance. We want safe work practices to become "habit strength," rather than suffering the consequences of workers placing themselves "at-risk." Review Figure 6-4 to see an example of a behavioral observation form that might be used in a facility handling chemicals.

Safety Programs

We want facility programs and policies that meet or exceed regulatory requirements (Figure 10-1 on page # and Figure 10-4 on page #). The objective is to establish a baseline of activity that promotes safety performance. Tip: write internal standards in the same outline/format as the regulatory standard. Add guidelines, procedures, exceptions, etc. under the category of "comments" or "implementation guidelines." This approach will greatly simplify writing policy and implementing programs.

Make sure the programs are relevant to the hazards and risk evaluation (Step 5). For example, ask "How can we take advantage of our existing hazard communication training for ammonia to jump start our Process Safety Management (PSM) program?" The answer to this question will go a long way in defining documentation needs for chemical process safety information, a key part of PSM. Another "trigger-question" to use to define goals and objectives is "How do we integrate PSM into release prevention elements of risk management planning (RMP) and pollution control?"

Learning and Growth

The goal here is to build employee knowledge and skills in order to remain competitive in a global market. From Chapter 9, we learned how big an influence safety and health training can have on incident prevention/reduction. A review of Chapter 9 will reinforce the steps to building a strong learning and growth infrastructure:

1. Assessment of training needs and planning
2. Select initial training programs
3. Develop and field test programs
4. Implement training and assess outcomes

A strong training program follows written specifications as to content and delivery medium, quality of instruction, and methods to assess trainee understanding.

Annual Safety and Health Plan

Too often, facilities do not have an annual safety and health plan. Without listing accomplishments and targeting measurable objectives each year, we have no basis for continuous improvement (Figure 10-3).

Organization is a key component of any safety and health plan. List accomplishments from the previous year in the same format as objectives from that year; then plan the objectives for the next year. It is much easier to measure progress when objectives and accomplishments are expressed in the same manner from year to year.

Figure 10-3 lists accomplishments and objectives in seven categories. The categories and order are used to convey a general sense of priority. For example, Ethanol 2000 management wanted to emphasize training, hazard identification and control, and safety policy. As in the example, each department should concentrate on hazard identification and control (item 2). Chemical operations should be targeted for HAZCOM and PSM training.

Note also that in the example, members of the safety committee (item 4) are hourly workers. Volunteers for the action teams come from the safety

---

2001 Safety and Health Accomplishments

1. Safety and Health Training
   Established site-specific training systems for hazard communication and process safety management.

2. Hazard Identification and Control
   Overall Plant
   Set up system for weekly department and monthly plant inspections.
   Purchased additional safety cabinets, emergency eyewash and first aid kits.
   Completed plant fire protection and deluge system inspection.
   Completed fire extinguisher floor plan.
   Completed noise monitoring.
   Commodities
   Used Job Safety Analysis to document rail car spotting procedures, identify hazards & safeguards, and serve as the basis for training.
   Repaired rope bin and electronic level indicators.
   Operations
   Installed new bulk sulfuric acid system.
   Safety shower installed in acid unloading area.
   Eyewash and safety shower installed for bulk ammonia system.

3. Policy and Enforcement
   Drafted Safety Management System (SMS) that includes:
   a. Management Values, Vision And Commitment
   b. Organization For Safety
      i. Safety Committee
      ii. Safety Staff
      iii. Safety Process Action Teams
   c. Line Accountability And Responsibility
   d. Safety Review And Improvement Process
   e. Training

4. Safety Committee Activities
   Reorganized monthly safety meetings and membership.

5. Employee Action Team Activities
   Established safety, health and environmental action teams for:
   a. Education andx training
   b. Health and environmental/PPE
   c. Rules and safe work practices
   d. Inspections/housekeeping and audits
   e. Fire and emergency preparedness
   f. Accident and incident investigation
   g. Process safety management
   h. Security
   i. Information sharing

6. Recordkeeping
   Appointed OSHA record keeper and documented new requirements for 2002.

Figure 10-3 Example of Safety and Health Plan

7. Federal and State Legal Requirements
   Setup file in safety library to document requirements and related plant programs. Completed the following activities:
   a. Process Safety Management
      i. Conducted PHA for new bulk sulfuric acid process and drafted management of change policy.
   b. Personal Protective Equipment
      i. Completed hazard assessments for all departments and drafted mandatory eye protection policy.
      ii. Ordered and received fall protection equipment for in-house and confined space entry applications.
   c. Lockout-Tagout
      i. Drafted SOP for ammonia system. Purchased locks for authorized personnel.
      ii. Drafted content and seeking quotation for on-site training.
   d. Confined Space Entry
      i. Drafted content and seeking quotation for on-site training.
   e. Portable Fire Extinguishers
      i. Drafted training content and seeking quotation for on-site exercise.
   f. Hazard Communications
      i. Completed multi-media and instructor-led sessions for all employees.
      ii. Determined requirements for electronic MSDS system.
      iii. Implementing use of NFPA labeling system.

2002 Safety and Health Objectives

1. Safety and Health Training
   Implement training for fire extinguisher, lockout/tagout and confined space entry training.
   Complete training for process safety management.
2. Hazard Identification and Control
   Overall Plant
   Follow-up with management staff to reinforce weekly and monthly inspections.
   Establish accident investigation procedure and action team. Select and investigate/document significant accidents, near-accidents, incidents and potential hazards.
   Commodities
   Complete annual refresher training. Continue to reinforce Job Safety Analysis and awareness of icing on rail cars and grain bins. Complete rail car capital project including purchase of fall protection equipment.
   Operations
   Finalize options for monitoring indoor air quality, improving ventilation and need for respiratory protection.
   Prioritize hazardous and non-routine SOPs for written procedures and/ or Job Safety Analyses (JSAs).

Figure 10-3 Example of Safety and Health Plan (continued)

CO_2
   Prioritize hazardous and non-routine SOPs for written procedures and/ or JSAs.
Maintenance
   Draft plan to complete lockout-tagout procedures for grains, operations, boiler/water and CO_2.
Boiler-Water
   Prioritize hazardous and non-routine SOPs for written procedures and/ or JSAs.
Laboratory
   Complete electronic MSDS database project.

3. Policy and Enforcement
   Continue to implement Safety Management System (SMS), including communication plan for all employees. Implement and monitor compliance to eye and face protection policy. Establish and implement hearing protection policy.

4. Safety Committee Activities
   Select and schedule managers to attend monthly meetings.

5. Safety Process Action Teams
   Develop plan to increase participation.

6. Record keeping
   Post 2001 OSHA 200 log (delete names) for month of February. Set up log and record keeping system for 2002.

7. OSHA Requirements- continue to document legal requirements and plant programs for Process Safety Management, Risk Management Planning, Personal Protective Equipment, Lockout-Tagout, Confined Space Entry, Portable Fire Extinguishers, and Hazard Communications.

Figure 10-3 Example of Safety and Health Plan (continued)

committee and co-opted workers recruited for specific tasks. The action team objectives support the plant written program (item 3- Safety Management System).

Federal and state legal requirements and the corresponding plant programs are documented in a plant safety library. The library should contain (minimally) the OSHA 1910 code books, three volumes of the National Safety Council's Accident Prevention Manual, and a file cabinet. From a functional standpoint, this approach has several benefits. First, the library is a source of reference material to all plant personnel. In addition, reading the file gives an actual account of what actions are taken. Thus it is easier to measure how the practice matches the policy. Finally, when OSHA makes a visit, the file eliminates the need to search multiple locations to find the documentation requested.

Use Figure 10-3 as an example from which the plant will select and add the priorities they consider important. Many of the items will come from monthly safety committee meetings. Others are system-related—for example implementation of the safety management system and increasing participation in action teams.

Step 2: Appoint an EHS coordinator(s) and provide adequate staffing and resources.

Write a job description and choose a safety coordinator. Provide qualified staffing to support safety, health, and environmental functions. Direct the safety staff's main activities toward hazard evaluation and control.

Too often, management will appoint an experienced technician or production supervisor to be the sole safety coordinator in the plant. This person (actually a support person) and the facility will flounder until he or she gets professional guidance, or until the person has had adequate time to climb the steep learning curve necessary to be effective.

The best approach is to staff the facility with professional and technical support that is based on facility size and level of hazards-risk factors. Considerations include the number of employees to train, number and complexity of OSHA PSM-covered processes, and the significance of personnel and contractor safety issues.

For example, if the facility is small (less than 200 employees), there are only one or two OSHA-covered PSM processes, and personnel/contractor safety concerns are few, then a single professional safety coordinator is usually sufficient to do the work. The person should have a science background and/or at least 2 years of safety experience. To continue professional development, the coordinator should have, or be seeking, safety technician status.

As the size of the facility increases, a safety professional and support staff is needed. If the number of employees to train is in the 200–500 range and there are several (3 or more) covered PSM processes, consider hiring a full-time professional with a BS degree in science and at least 5 years of safety experience. The support person may work on safety up to full time. He or she ought to be seeking associate safety professional status (ASP).

For large facilities (more than 500), the safety person should be certified or working toward the certified safety professional (CSP) status. The facility will likely need more than one support person. If there is a significant amount of chemical handling, processing, and exposure, an on-site or assigned staff industrial hygienist will probably be needed to handle employee health functions and other medical elements of the program. Appoint an environmental coordinator to work on emission and hazardous waste control activities. An ergonomist (person trained in ergonomic principles to address material handling issues and musculoskeletal disorders (MSDs)) may also be needed.

Facilities always have many needs and are competing for scarce resources. In some cases, hiring a safety professional may seem like an unaffordable option. Contact OSHA consultation, the American Society of Safety Engineers, or the American Institute of Chemical Engineers for a recommendation. Or consider contracting for a qualified safety consultant to work with management and the inexperienced safety coordinator until other choices become workable.

Companies that have a number of plants should prioritize locations in need of professional EHS help. Toward this objective, there are distinct benefits to setting up a company OJT training program for new graduates in safety and/or industrial hygiene. These people can be placed according to the priority schedule and assigned to build their skills and the EHS function while working with technical support people in the plant. Plant safety personnel (professional and technical support) should have their personal training needs identified as part of their development plan. Staffing and training efforts need to be focused on specific plant programs and priorities such as incident investigation, confined space entry, process safety management, etc.

Examples

One of my clients called me in to do a PHA study on an OSHA-covered process. The job of safety coordinator was a temporary, part-time position for the plant personnel manager. Following the study, the facility hired a full-time safety coordinator that had no background in Process Safety. It was my job to coach the safety coordinator how to work with line management until he was self-trained in PSM.

One company's program (called Optimized Operations for Safety) is an example of moving new graduates (designated O2 engineers) through a staff function for six months to one year before getting a plant assignment. The graduates have majors in safety from accredited institutions and, depending on the sponsoring division, some have industrial hygiene training. The plants benefit from having experienced, professional capabilities immediately. The O2 engineers benefit by having a career path ahead of them in safety and health.

Step 3: Define line-management accountability for safety. Provide performance measures and indicators.

Remember from the values discussion above and Chapter 8 that line-management is responsible for day-to-day safety and health. Operational managers need to lead the safety effort just as they do productivity, quality, cost, and personnel relations. You will want to position the safety coordinator and other EHS staff into a small group that is driven by the facility manager and line management. For example, Ethanol 2000 wisely installed a safety steering committee to drive the program, composed of the general manager and managers for technology, operations, maintenance, and accounting (EHS recordkeeping). Then line management helps a number of action teams define hazard reduction and control proposals sent to the safety steering committee for approval.

Establish performance measures and indicators for the facility in a careful fashion. After-the-fact results such as lost time injuries, restricted work cases, OSHA recordables, spills, etc. are a necessary reality, but provide little direct opportunity for improvement. Pay attention to these numbers, but install preventive, upstream performance indicators and control measures as well. Concentrate on activities such as safety audits, safety devices installed, PHAs and other hazard control sessions held as scheduled, air sampling programs, etc. (Table 7-1). Measure and reinforce activities like these wherever possible and especially when safety performance is a long way from the goal. Ask each department head in your organization to brainstorm 2-3 indicators of safety performance and continuous improvement.

Step 4: Initiate opportunities for employee involvement.

In order to provide identification and ownership, employee involvement is essential to the EHS improvement process. Workers that are involved in setting standards for safe work practices, participate in safety committee meetings, and have individual safety-related assignments are much more likely to have a positive safety attitude than those not involved in these activities.

Ethanol 2000's hourly employees serve on action teams listed in Figure 10-3 (Accomplishments, item 5).

Action team status updates are given each month during the safety committee meeting.

Think of how your plant safety program could be revitalized if employees:

1. Participated in design and safety reviews prior to accepting new chemical process equipment from the supplier
2. Helped the safety department select safety glasses, respirators, and other PPE based on the hazards of the job
3. Worked on chemical spill-release problem solving teams
4. Volunteered to conduct behavioral observations
5. Served on incident-near miss-potential hazard investigation team
6. Consistently went beyond pointing out safety problems to make suggestions for corrective action

Employee involvement will not last without positive reinforcement. Review the list in Figure 6-5 and add positive reinforcers to your measurement and feedback program.

Self-Directed Work Teams

Self-directed work teams are increasingly found in today's industrial and government facilities. One could say that this is complete employee involvement in safety, but there are reasons to be cautious when initially assigning safety responsibilities to self-directed work teams.

In the early 90s, many facilities, including chemical plants, "right-sized" by replacing retiring supervisors with lead operators on self-directed work teams. Initially, this change caused safety problems because it seemed that no one was responsible for safety. Then safety training was changed to include lead operators and in some cases the entire work team. Because management was committed to safety and participated in the training, labor-management trust was strengthened and self-motivation to work safely improved.

Taking Safety Home

This is another tip to improve awareness and employee involvement at work. Employees are generally a lot safer at work than at home. Injuries off-the-job affect job performance as much as if the worker was injured at the plant. Therefore, safety and line managers should also encourage safety at home. For example, "safe driving; handling household chemicals; home security issues; ergonomic and MSD problems associated with hobbies and other off-the-job activities; slips, trips and falls; use of PPE; and ladder safety" are all good areas for encouragement.[4]

Step 5: Develop a system of hazard-risk evaluation and control measures.

Hazard-risk evaluation and control is at the heart of any facility safety management system. Note that this system along with Step 6 (Drafting policies and procedures) needs to be integrated with applicable regulatory requirements (Figure 10-1 on page 297 and Figure 10-4 on page 317).

There are several sub-systems to install and recognize as key hazard-risk control measures:

A. Establish a Self-Inspection Program.

Inspections are treated in Chapter 5 as the primary way to assure safe work conditions related to personnel safety. Self-inspections for the total plant and individual departments are an important part of a family of inspection categories that include formal audits, regulatory inspections, special technical inspections, and loss prevention surveys. A comparison of the scope and degree of employee involvement in self-inspections usually separates a good safety program with one that has a lot of room for improvement. Figures 5-1 and 5-2 will help the reader establish specific items to inspect within the facility. Figure 5-3 is a good approach to corrective action. Note that self-inspection and follow-up is another area where ownership is a key factor. Allow participants to come up with their own list, even if incomplete, and add categories on a priority basis. Seek help from a professional to aid inexperienced inspectors understand what to look for. Use the Safety Precedence Sequence (Figure 4-6) and employ engineering controls whenever possible to accomplish corrective action.

Examples

Consider the ammonia leak, discovered during an inspection and described in Figure 5-3. The corrective action has two parts. The immediate corrective action is to repair the flange. The second fix, higher on the precedence sequence, adds a remote shutoff valve closer to the ammonia tank.

In the second example, say that we are inspecting a dipping and coating operation and turn to 29 CFR 1910.124 for guidance. Paragraph (g) describes in very specific and rather unsophisticated terms what hygiene options may be used in the event that a splash lands on a worker's face and body. 29 CFR 1910.151 (Medical Services and First Aid) states only that "where the eyes or body of any person may be exposed to injurious corrosive materials, suitable facilities for quick drenching or flushing of the eyes and body shall be provided within the work area for immediate emergency use."

In this case, it is helpful to have a facility or corporate standard that provides design requirements for eyewash and safety shower units. For example, you should indicate locations for the safety shower like "no more than 50 feet away or require more than 10 seconds travel from the hazard." For more corrosive materials like strong acids or bases, the shower should be closer, say within 25 feet. Your standard should cover features such as shower head distance from the floor, size of the spray pattern, gallon per minute capacity (e.g., suggest minimum of 25 gpm of tempered water for 15 to 20 minutes). Consult the chemical's MSDS for specific guidelines. It is common to read "flush with plenty of water for 15 minutes."

B. Initiate or Extend the HAZCOM and Other Personnel Safety Program to Comply with Process Safety Management and Risk Management Planning Requirements.

Start by reviewing the regulatory requirements for the two OSHA standards (Figures 8-2 and 8-4). The six mandatory parts of Hazard communication are items (b) through (i) and apply to *personnel exposures* related to chemical safety. Note the requirement for a written program. The author discourages safety coordinators from writing this program from scratch. Take advantage of services offered by many consultants or purchase a template[5] and be sure to customize the program to your situation. Remember that a written program means nothing unless you do what you say you are doing out on the floor. Pay particular attention to labeling, material safety data sheets, and employee information and training. In addition to a paper list, you should keep an electronic list of covered chemicals (1910 Subpart Z—Toxic and Hazardous Substances). Most air contaminants are listed by OSHA along with 8 hour, time-weighted averaged permissible exposure limits (PELs). Determine how average and peak exposures relate to the limits and search the current version of the Guide to Occupational Exposure Values, compiled by the American Congress of Government Industrial Hygienists (ACGIH®).[6] This is important, for the ACGIH sometimes lists more conservative exposure values for chemicals than OSHA.

Next, review the comprehensive (14 elements, 1910.119(c) through (p)) requirements for process safety management (PSM). The purpose of PSM is to prevent occurrence of catastrophic *process-related* incidents, namely large-scale fires, explosions, and toxic releases. The chemical list (1910.119 Appendix A) is different than that for either HAZCOM or RMP, but there is a good deal of overlap.

Recognize that much of the PSM activity is for the facility to develop and maintain current process safety information for the life of the covered process. Most of this can be accomplished through use of process hazards analysis (PHAs) studies and carefully documenting results of the follow-up. Keep an up-to-date block or process flow diagram (it is acceptable to temporarily red-line changes). Remind your process safety coordinator/committee to maintain the so-called "engine" of process safety management (Figure 8-5) and to work diligently on mechanical integrity (which is mainly the preventive maintenance element for process safety management).

Process Safety Management is the accident prevention plan for EPA's Risk Management Plan (40 CFR Part 68, Figure 8-8). Subpart F contains the list of regulated chemicals. Note that the threshold quantities for RMP are sometimes different than those for OSHA's PSM. In addition, EPA's controversial threshold determination (Subpart F 68.115) is different for toxic materials than the permissible exposure values governing HAZCOM. Originally, the term defining the likely offsite impact was *toxic end point*, now commonly referred to as *vulnerability zones*. Table 8-5 provides a comparison of key parameters for common PSM and RMP covered chemicals.

Chapter 8 (Implementing the Risk Management Program) provides an eleven-step procedure for drafting and submitting the RMP.

Review the other regulatory requirements in Figure 10-1 (see Chapter 8 for more detail) and the Code of Federal Registry (CFR) for items applicable to your facility. This exercise may well persuade you to seek additional resources in the form of a consultant and/or corporate EHS functions.

C. Establish or extend the Preventive Maintenance (PM) Program.

The ideal model for an effective chemical safety preventive maintenance program is laid out in OSHA's PSM standard under Mechanical Integrity. Even if all facilities and equipment are not covered by the standard, the five Mechanical Integrity requirements represent, at the very least, good industry practice:

1. Written procedures must be in place for maintaining the functionality of critical process equipment, including pressure vessels and storage tanks, piping systems, relief and vent systems, emergency shutdown systems, controls and monitoring devices, alarms and interlocks, pumps, compressors, filters, and fired heaters. Concentrate here on the practical aspects of how the maintenance function must operate. For example, it is better to expend resources toward training than to expend a huge effort to document each PM procedure in detail.

2. There must be an established training program for maintenance people. The training is to include an overview of the process and associated hazards as well as how to conduct applicable safety procedures. Table 9-2 provides a guide to training for five mechanical integrity inspector levels along with typical tests and procedures.

3. Equipment inspections and tests must be done according to recognized and generally accepted engineering practices. Deficiencies found must be corrected prior to use or in a safe and timely manner given that temporary safe operation is assured. Acceptable limits for the tests are to be defined by the process safety information for the equipment. Inspection and test frequency often becomes an issue due to efforts to maintain cost control. The OSHA standard requires recommendations from equipment manufacturers, following "good engineering practices and more frequently if determined to be necessary by prior operating experience."[7] Although these guidelines do not provide a definite frequency, we suggest you consult with design and maintenance engineers familiar with the chemical service and materials of construction involved. This type of information, balancing preventive maintenance, and equipment failure/downtime costs should give you the answers you need. Furthermore, training your own inspectors will provide flexibility in conducting the tests, allow you to build an experience base, and enable you to establish the frequency of inspections.

4. Documentation of the inspections and tests must be maintained (see 29 CFR 1910.119(j)(4)(iv)) for details. Maintain manually written records and consider purchase of commercially available software. Do not attempt to force installation of electronic systems designed to schedule maintenance if they are will not adequately document inspections and tests.

5. OSHA calls for a quality assurance system to assure that any new equipment as fabricated is suitable for the intended process application. Review Chapter 3 (Function of Codes and Standards) to reinforce using design standards as a basis for setting equipment specifications for materials, fabrication, and installation. Conduct inspections and checks during the fabrication and installation to be sure equipment will function according to the design specification and the manufacturer's instructions. Also assure that maintenance materials, spare parts, and components are readily available and suitable for use. Here you will need to implement management of change procedures so that safety is not compromised when replacing or maintaining equipment. See Figures 5-6 and 5-7 plus the discussion for more details.

D. Evaluate Plant Operating and Maintenance Procedures and Prioritize/Document Those Judged as Hazardous.

Review Chapter 6 (Safe Work Practices) and the discussion of Job Safety Analysis and use of Behavioral Observation methods. As was recommended above for maintenance, design the training such that documentation is minimized. Integrate procedure writing and implementation of quality systems such as ISO 9000 and 14000 wherever possible.

E. Institute Use of the Safety Precedence Diagram (Figure 4-6) and Strengthen the Facility Personal Protective Equipment (PPE) Program.

Figure 4-6 is presented in the context of accident/incident investigation but applies broadly to preventive and corrective action. PPE is too often the first choice when it should be the last resort in preventing worker exposure to a chemical or physical hazard. Use of engineering controls to reduce or eliminate the hazard should always be the goal of equipment suppliers and facility personnel.

Start the facility PPE program analysis with a hazard assessment as required by 29 CFR 1910.132(d). See Subpart I, Appendix B for hazard assessment guidelines and PPE selection. Figure 8-3 is an example of a hazard assessment specifying typical maintenance operations, hazard categories, and appropriate PPE.

Choice of the right gloves for handling diverse chemical operations (acids, bases, solvents, toxic and flammable materials, etc.) can be complex. Even if you have industrial hygiene resources, consider purchase of glove selection software[8] or seek sources of glove-chemical resistance information.[9]

You can select appropriate PPE from the hazard assessments throughout the plant and incorporate the PPE into written procedures using Job Safety Analysis (Figure 6-2).

F. Evaluate the EHS Impact of All New Facilities, Materials, Equipment, and Processes Prior To Startup.

This ongoing activity is best accomplished by holding design, project, and pre-start safety reviews at key stages during the project. Safety needs to be incorporated into the design as early in the project process as possible. New equipment should pass a final safety and operability review before it is accepted by the facility and leaves the supplier's shop. OSHA's PSM standard, once again, is the model for chemical processing industries. See Chapter 8 and 29 CFR 1910.119(i) for pre-start requirements. Also use the concepts presented in Figure 3-4 (Multiple Independent Safety Layers) at the beginning of a project. Deal with machine guarding and lockout-tagout issues according to your company/facility standard or 29 CFR 1910.211–1910.219 and 1910.147, respectively. Note that the general requirement description for all machines is quite brief. If machine guarding and lockout-tagout resources are limited, we suggest you contact an automation/controls consultant with expertise in your type of machine and/or an equipment supplier.[10]

Finding Resource Information for Hazard Control

Having immediate access to sources of information and equipment is important to the success of the safety function. Such references have been provided throughout the book. The listing of consensus organizations and standards (Figure 8-10) is a good example. Also see Table 10-1 (Sources of EHS Information).

Table 10-1 Sources of Environmental, Health, and Safety Information

American Chemistry Council www.cmahq.com	See Responsible Care initiatives, news and media, workshops and conferences, research and testing, federal corner, site security guidelines.
American Institute of Chemical Engineers www.ai.che.org	See conferences and programming, publications, member activity groups, education and training, industry focus, resources, and government relations.
Center for Chemical Process Safety www.ai.che.org/ccps	See ProSmart® PSM measurement software; process safety alerts, courses and incident database; equipment reliability (PERD); CCPS professional assistance directory; & safety and chemical engineering education program (SACHE).
American Petroleum Institute www.api.org	See clean air, clean water, waste management & recycling, global climate change, health, safety, and clean fuels.
American Society of Safety Engineers www.asse.org	See disaster checklist, SHE projects, practice specialties, professional development & education, conferences & expositions, government affairs.
Air and Water Waste Management Association www.awma.org	See events, publications, training and resources; public outreach, government agencies directory, job listings.
The Chlorine Institute www.cl2.com	See what's new, rmp info, World Chlorine Council, annual report, safety training, bio-terrorism.
Environmental Defense www.environmentaldefense.org	See scorecard, climate change, action center, biodiversity, health and oceans.
Environmental Protection Agency www.epa.gov Chemical Emergency Preparedness and Prevention Office www.epa.gov/swercepp	See prevention, preparedness, response, international counter-terrorism; stakeholders, resources.
Federal Emergency Management Agency www.fema.gov	See hazards and assistance, disaster communities, emergency personnel, education & training, news media, regions.
National Safety Council www.nsc.org	See topics and highlights, service and news, training, new offerings.
Occupational Safety and Health Administration www.osha.gov	See compliance assistance, cooperative programs, newsroom, safety and health topics, statistics, international.

(continued on following page)

Table 10-1 Sources of Environmental, Health, and Safety Information (continued)

Paper, Allied-Industrial, Chemical & Energy Workers International Union www.paceunion.org	See calendar, downloads, emergency response team, featuring locals, links, worker resources, the pacesetter.
Public Interest Research Group www.pirg.org	See for the media, state PIRGs, news in Washington D.C. and across the country, Clean Air/Water Act.
Society of the Plastics Industry www.socplas.org	See business units, news and publications, calendar of events; about the industry, issues and public policy, outreach and education, business development.
Synthetic Organic Chemical Manufacturer's Association www.socma.com	See conferences and workshops, Responsible Care®, issues and advocacy, networking tools, products and services, site security, press room.
Department of Transportation Office of Hazardous Materials Safety www.hazmat.dot.gov	See rules and regulations, exemptions and approvals, training information, publications and reports, news and discussions, spills, emergency response guidebook, risk management, hazmat enforcement, other transportation links.
The Vinyl Institute www.vinyl.org	See automotive, toys, construction, medical, and electronics industries; environmental attributes of vinyl, news and publications center, related links.

Finally, we would be remiss without mentioning Best's Safety and Security Directory.[11] The two-volume directory is free and is an excellent addition to the plant safety library. Subjects include OSHA requirements, insurance loss control and consulting, PPE, environmental controls, employee health and hygiene, machine guarding and tool handling, hazardous and ordinary material handling, plant maintenance, electrical lighting, and fire safety and security. The 15 chapters each feature helpful information including an OSHA summary, self-inspection checklist, safety guidelines, training articles, product descriptions, and a buyers' guide.

Step 6: Draft EHS policies and procedures that meet or exceed regulatory requirements.

We want the work of safety, health, and environmental functions to be determined by facility policy, not by external regulatory requirements. We also see from Figure 10-1 that both steps 5 and 6 need to be linked to regulatory requirements. This linkage is shown conceptually and in more detail in Figure 10-4.

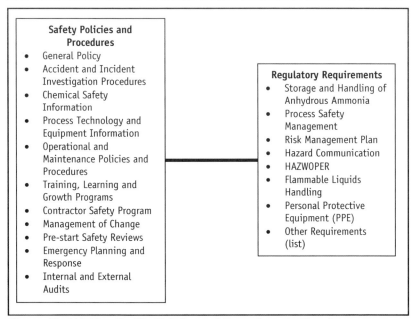

Safety Policies and Procedures
- General Policy
- Accident and Incident Investigation Procedures
- Chemical Safety Information
- Process Technology and Equipment Information
- Operational and Maintenance Policies and Procedures
- Training, Learning and Growth Programs
- Contractor Safety Program
- Management of Change
- Pre-start Safety Reviews
- Emergency Planning and Response
- Internal and External Audits

Regulatory Requirements
- Storage and Handling of Anhydrous Ammonia
- Process Safety Management
- Risk Management Plan
- Hazard Communication
- HAZWOPER
- Flammable Liquids Handling
- Personal Protective Equipment (PPE)
- Other Requirements (list)

Figure 10-4 Linking Facility Policies and Procedures to Regulatory Requirements

Management needs to develop plant EHS policies, procedures, practices, and hazard-risk control programs that meet and go beyond the minimum requirements of federal, state, and local regulations. Since organizational policies are no more *optional* than regulatory requirements, the policy should drive the continuous improvement process.

OSHA, EPA and DOT promulgate overlapping regulations. For example, OSHA's comprehensive Process Safety Management (PSM) program includes elements of the hazard communication standard (see Chapter 9). Also, PSM is the prevention program part of EPA's Risk Management Planning standard. Those involved in chemical safety management need to acquire and understand the applicable regulations in order to define safety policies and procedures in the plant safety manual (Figure 7-2).

Take advantage of regulatory language that is available in electronic form. Easily customized, written programs are commonly available.

There are also voluntary programs (such as Responsible Care) that are designed to comply with OSHA, EPA, and DOT regulations. See Chapter 11 for a description of this initiative.

Step 7: Initiate internal and external communication plans

If you have completed all the steps suggested to this point, you have made a considerable investment in safety, health, and environmental stewardship. The investment will be lost, however, without an internal and external communication plan. Well planned and executed, the communication plan can be the most enjoyable part of the entire improvement process.

Internal Communications

Employees have the most at stake since their safety and health (and often the safety and health of their families) is on the line every day. We know that attempts to improve employee attitudes through gimmicks, campaigns, memos, and handouts will surely fail. Management needs to be aware of and constantly work to eliminate the causes of negative attitudes about safety and health. In a good program, the entire organization works together to build and maintain a safe, healthy operation that protects the environment. Review the management behaviors provided in Figure 7-3. These and other management-employee activities will go a long way to demonstrate management's commitment to EHS stewardship.

Community Outreach

Clearly, the facility is not alone in the business of chemical safety management. There are many others (called stakeholders) who are affected or interested in what you do about chemical accident prevention and other chemical safety issues. Consider the following list of individuals and organizations that have a high interest in what you do:[7]

1. Employees
2. Suppliers and customers
3. Federal, state, and local regulators
4. Business, political, and social service communities
5. Regulatory or environmental activist groups
6. Major audience groups within each community
   - Educators, doctors, business leaders
   - Emergency responders
   - Environmental activists organizations
   - Local government leaders
   - Local business leaders

- Members of the media
- Neighbors (near)
- Retirees
- Social service leaders
- State government officials (elected and agency)

These groups (and individuals within the group) may have little interest in the facility EHS program. It is more likely that once prompted, many stakeholders will express a negative view, especially in the absence of objective information. The benefits of a strong internal and external communications program are many, but perhaps can be summed up with the statement, "If an incident occurs, you want the benefit of doubt."

Employees are the best ambassadors to the community. Neighbors and others in the community may not know much about safety at the facility but often will express strong opinions about noise, traffic, smoke, odor, and other complaints.

Implement the following steps in order to develop a community outreach program:[12]

1) Define your community.
2) Profile the community.
3) Establish your team's expectation.
4) Identify the challenges, roadblocks and opportunities for your team.
5) Plan actions that will help you get where you want to go.

Develop, communicate, and reinforce environmental, health, and safety messages that demonstrate your company's achievements and commitment to continuous improvement.

Community-Right-To-Know

As a result of EPCRA (Chapter 8), some estimated 66,000 facilities that handle what EPA calls "extremely hazardous substances" are required to inform workers and the public what could go wrong in chemical accidents. Plants are required to develop worst-case and most-likely accident scenarios that are pictured on a map. Once known, these worst-case circles, or "vulnerability zones," around a facility are apt to alarm the community. Appendix A and B provide the details of how to determine the vulnerability zone.

The benefits of community outreach in such a situation are threefold. First, it is an effective way to meet federal employee and community right-to-know requirements. In addition, the outreach effort builds trust and goodwill among community stakeholders. Finally, the goodwill generated by a strong outreach effort will reduce complaints about noise, odor, dust, and other issues that should not be ignored.

Be prepared to answer some tough questions[13] about chemicals you have on-site. For example you may be asked:

"What chemicals do you have over there that can hurt my family? How many people could be injured, killed or have to be evacuated inside that worst-case circle around your facility?

If there is a release, how will I get information to protect my family?"

Also, ask yourselves the following questions internally before they come from outside:

"How long will it take to find a leak?

How long might it be for us to decide to report a release to the state or county or call the fire department?

How long will it take for workers and neighbors to evacuate or shelter-in-place?

How can we improve site security and reduce potential sabotage?

Anticipate these and other questions and practice your answers. It is important to begin building and strengthening relationships in the community immediately. As Deborah Gelbach says "You want to make friends before you need them." If you wait until an incident or confrontation occurs, it is often too late to establish a good relationship, and it is unlikely that you will be provided the benefit of the doubt. Note that at the very least, the facility must comply with the requirements of employee and community right-to-know regulations. EPA's risk management regulation requires facilities to coordinate their plan with the community plan even if they rely on professional emergency responders.

Start building the relationship early; talk with local emergency organizations such as the fire department, county emergency services office, and the local emergency planning committee (LEPC). Conduct fire extinguisher exercises with local fire fighters. Provide 24-hour access to plant emergency telephone contacts, maps showing electrical power switches, and current copies of MSDS information. Although your facility may not have had any accidents requiring outside response, this type of cooperation increases emergency preparedness and provides a goodwill benefit.

Consider where your facility is located with respect to neighbors and communities in your area. Be concerned with geographical borders and political, business, and social services communities. Get to know the regulatory and environmental activist groups and their points of view. Remember that your employees and retirees are also stakeholders. These people can be a tremendous asset in getting your messages across to the public.

Internally, discuss the major issues that could impact the facility's community relationship. Consider labor problems, emissions/pollution, complaint and solution history, accidents and chemical release incidents, traffic, noise, and other issues that concern people. Keep a file of news items with which your team should be familiar. Maintain a file of complaints and compliments received by telephone.

Establish realistic expectations. Do not attempt to implement everything at once, but lay out priorities and tactics consistent with your resources. Hold a brainstorming session with your outreach team, listing goals and strategies for the coming year. Consider ideas such as:[14]

- Open House celebrations for EHS achievements
- Interviewing individuals and groups to get feedback from the community
- Supporting important social causes shared by residents
- Demonstrating that the facility complies with local, state, and federal regulations
- Celebrating a long-term relationship with the community
- Supporting emergency training exercises at your facility

Be realistic about challenges, roadblocks, and opportunities. Challenges can be overcome with a good plan and follow-through. Getting around roadblocks may require additional resources or a change in priorities. Include considerations such as budgets, lack of community contact, top management disinterest, and community leader distrust. Plan outreach opportunities such as community feedback meetings, supporting community improvement projects, special days for community celebrations, and breakfast meetings between the mayor and plant manager.

Step 8: Provide training and opportunities for learning and growth.

Facilities often do not give enough attention to choice of training materials, quality of content, and meaningful follow-up to determine trainee understanding. Supervisors may not be trained to teach and coach employees to work more safely.

Chapter 9 stressed the importance of employees relating safety and health training directly to their jobs. It is therefore important to integrate product-process training with EHS programs wherever possible.

Establish a process in order to install a sound EHS training infrastructure:

1. Determine employee knowledge and skills needed
2. Draft key operating and maintenance procedures on which to base the training content
3. Draft a training specification
4. Select and tailor the training program to meet your facility's needs
5. Provide a system to train the trainer
6. Develop and maintain a training documentation system that includes a means to verify trainee understanding
7. Provide initial training and refresher sessions that again reflect individual employee needs

Choose and evaluate off-the-shelf training programs carefully. Always start with simple, low-cost media such as Power Point® slides before investing in more expensive, more complex enhancements like multi-media Computer-Based Training. Take advantage of opportunities to evaluate and use free training packages available from OSHA (Figure 10-5).[15]

• Introduction	• Hand and Powered Tools
• OSHA Procedures	• Emergency Response
• Safety and Health Program	• Lockout- Tagout
• Regulatory Agencies	• Confined Space Hazards
• Occupational Health Professionals	• Permit Spaces
• Industrial Hygiene Safety	• Electrical
• Accident Investigation	• Walking Surfaces
• Personal Protective Equipment	• Hazard Communication
• Flammable Liquids	

Figure 10-5 OSHA's Small Business Outreach Training Topics

Note that OSHA provides downloadable, overhead masters in either HTML or PDF formats and in many cases includes student handouts. Be sure to make the training site-specific.

All participants need not take the same version of the five compulsory topics in Table 10-2. Further, a topic such as hazard communications should be changed to fit the needs of individual departments. For example, chemical operations and the laboratory need to cover all chemicals handled by the facility. The scope of HAZCOM may be more limited for other departments.

Note also that process safety training is much more comprehensive for operations than for the maintenance function and management overview of the topic. Generic Power Point® slides or multi-media presentations may be used for the management overview. However, operators should be trained in the control room, using primarily the distributed control system and material from the facility operations manual. Refresher training for operators is conducted over a three-year period, meeting the OSHA PSM requirement.

Refer to Table 9-1 as an example of generic chemical operator training. Employees also need to learn and grow from a variety of situations beyond the classroom. Reinforce required training topics such as lockout-tagout, confined-space entry with simple on-the-job exercises.

Make sure that during or prior to the training, employees become familiar with use of safety and personal protective equipment. For example, using an emergency squad from outside the facility for hands-on confined space entry training can generate interest and excitement as well as effectively demonstrate the proper use of PPE. The equipment furnished might include respirators, harnesses, air monitors, and their instructions for use. Often, this will prompt a discussion among trainees about equipment they do or do not have at their facility. When trainees do not know how to use certain equipment, it quickly becomes apparent. Such training practices can also provoke animated conversations such as "We need this or that," or "We'd better get batteries for our air monitor and read the instruction manual before we start our own training."

Design the scope and format for training documentation to meet the needs of the organization and facility. OSHA requires only that you record the topic, employee's name, date of training, and test score or other means to verify that the employee understood the training. Keep this level of

Table 10-2 Safety Training/Orientation Program—Site-Specific Training

	Hazard Comm.	Process Safety	Fire Prevention	Lockout-Tagout	C. Space Entry	Notes
Primary participants	Operations Boiler and Water Laboratory Carbon Dioxide Commodities Janitorial Maintenance Administration (overview) Laboratory	Operations Maintenance Administration (overview)	Operations Administration (overview) Carbon Dioxide Commodities Boiler and Water Janitorial Laboratory Maintenance	Maintenance Operations Boiler and Water Commodities Carbon Dioxide Contractors (orientation) Administration (overview)	Maintenance Operations Boiler and Water Commodities Carbon Dioxide Contractors (orientation) Administration (overview)	
Means of training	Customized MM Power Point® slides Site-specific follow up On-the-job-training (OJT)	MM Power Point® slides Operations manual OJT	MM Power Point® slides Operations manual OJT Fire extinguisher and hose drills (with Windom FD?)	MM Power Point® slides Site-specific follow-up OJT	MM Power Point® slides Site-specific follow-up OJT	
Site-specific information	Chemical inventory Emergency contact list MSDS information Hazards of non-routine tasks Location of eye wash and safety showers	Process Safety Information -chemicals -process condition -safety system Management of Change Mechanical Integrity Hot work permit	Location and use of fire ext. (drill) Evacuation drill Specific ignition sources Working in classified areas Labeling and warning systems	Authorized and affected ee's LOTO system LOTO locations SOP 90- Grains SOP 100- Ammonia SOP 110- Operations SOP 120- Boiler/Water SOP 130- Carbon Dioxide SOP 140- other (specify)	Authorized and affected ee's Facilities and equipment requiring entry Specific hazards Air testing equipment Air test and monitoring procedure Entry and work procedure Entry permit and approvals	Lockout SOPs will include line-breaking procedures

documentation for the employee file and back it up with records filed by topic. Include, at a minimum, a listing of necessary skills for each employee group, course outlines and lesson plans, and results of a periodic review of the training. Strongly consider designing and purchasing documentation software that meets facility and central organizational needs.

Step 9: Strengthen OSHA recordkeeping and develop a functional management system for safety, health, and environmental information.

Consider the growth in recent years of environmental regulations and quality initiatives such as EPA's Risk Management Program, Title V operating permits, ISO 9000, and ISO 14000 certification programs. Safety, health, and environmental management systems have changed the perception from something companies consider "nice to have" to something "worth considering."[16]

An important consideration is to assure that the chemical safety management system is as effective as other business priorities. To that end, facilities need an efficient recordkeeping and effective management information system (MIS) for safety, health, and environmental activities.

Designing and installing an EHS management information system is an important undertaking. Give careful attention to the following issues:

1. What is the scope and relative costs for the recordkeeping and management information system we need? What do we need to link business-related objectives/measures and related EHS activities?

2. What will it take to better share data, resources, lessons-learned, and other information with other plants and staff groups within the company? What do we need to link internal and external information sources?

3. How do we acquire the information needed to satisfy the regulatory (OSHA, EPA, and DOT) requirements and still meet our internal and external needs?

4. Will we be able to match and/or use forms and systems already in use?

5. How do we get the design and training resources necessary to implement and maintain an EHS management information system? Will we have the ability to change technical support for the system if problems develop?

OSHA s Form 300

Log of Work-Related Injuries and Illnesses

Attention: This form contains information relating to employee health and must be used in a manner that protects the confidentiality of employees to the extent possible while the information is being used for occupational safety and health purposes.

Year 20___ ___

U.S. Department of Labor
Occupational Safety and Health Administration

Form approved OMB no. 1218-0176

You must record information about every work-related death and about every work-related injury or illness that involves loss of consciousness, restricted work activity or job transfer, days away from work, or medical treatment beyond first aid. You must also record significant work-related injuries and illnesses that are diagnosed by a physician or licensed health care professional. You must also record work-related injuries and illnesses that meet any of the specific recording criteria listed in 29 CFR Part 1904.8 through 1904.12. Feel free to use two lines for a single case if you need to. You must complete an Injury and Illness Incident Report (OSHA Form 301) or equivalent form for each injury or illness recorded on this form. If you're not sure whether a case is recordable, call your local OSHA office for help.

Establishment name _____

City _____ State _____

Identify the person

Describe the case

Classify the case

(A) Case no.

(B) Employee's name

(C) Job title (e.g., Welder)

(D) Date of injury or onset of illness

(E) Where the event occurred (e.g., Loading dock north end)

(F) Describe injury or illness, parts of body affected, and object/substance that directly injured or made person ill (e.g., Second degree burns on right forearm from acetylene torch)

Using these four categories, check ONLY the most serious result for each case:

(G) Death
(H) Days away from work
(I) Remained at work — Job transfer or restriction
(J) Remained at work — Other recordable cases

Enter the number of days the injured or ill worker was:

(K) On job transfer or restriction — days
(L) Away from work — days

Check the "Injury" column or choose one type of illness:

(M)
(1) Injury
(2) Skin disorder
(3) Respiratory condition
(4) Poisoning
(5) All other illnesses

month/day

Page totals

Be sure to transfer these totals to the Summary page (Form 300A) before you post it.

Public reporting burden for this collection of information is estimated to average 14 minutes per response, including time to review the instructions, search and gather the data needed, and complete and review the collection of information. Persons are not required to respond to the collection of information unless it displays a currently valid OMB control number. If you have any comments about these estimates or any other aspects of this data collection, contact: US Department of Labor, OSHA Office of Statistics, Room N-3644, 200 Constitution Avenue, NW, Washington, DC 20210. Do not send the completed forms to this office.

Page ___ of ___

Figure 10-6 OSHA's Form 300—Log of Work-Related Injuries and Illnesses

6. How easily can changes be made to the system after initial implementation?

7. What are our needs for information security at various levels within the facility and throughout the entire organization?

Thoroughly discuss these and other challenges that may be unique to your organization.

Improving the Facility OSHA Recordkeeping System

At the beginning of 2002, OSHA established a new and simplified recordkeeping system for tracking recordable injuries and illnesses. Labor Secretary Elaine Chao announced the new standard by saying, "This rule is a big step forward in making workplaces safer for employees. It is written in plain language and simplifies the employer's decision-making process."[17] There are many changes from the old OSHA 200 log system—most are major improvements. For example, the new form 300 no longer distinguishes between work-related injuries and illnesses (Figure 10-6).[18]

Download a package from the OSHA web site that contains all you will need to set up and implement your new system:

- An overview and general instructions for completing the forms
- An example of how to fill out the OSHA 300 log
- Several pages of the OSHA 300 log
- An injury and illness summary form that can be posted at the end of the year
- A worksheet to help record keepers complete the summary
- The OSHA 301 Injury and Illness Report (required- provides much of the same information as Figure 4-3)

Record keepers need to be alert to changes affecting recording chemical-related injuries and illnesses. The new 300 form has an injury column that may be used to record instantaneous chemical exposures and three illness columns for skin disorders (e.g. dermatitis from chronic chemical exposure), respiratory conditions, and poisoning. Record keepers are advised not to throw away the old BLS booklet.

Finally, you should take advantage of the new feature that requires company executives to certify the 300A summary form, indicating they have examined the document for accuracy. This is another opportunity to promote top management awareness, thorough accident investigation, and hazard identification/control.

Developing a Functional Safety, Health, and Environmental Management Information System

Large facilities often require large, robust, and flexible management information systems. Smaller organizations may not even need an electronic system. However, large or small, every facility needs a management information system of some kind to help them track safety, health, and environmental information.

Selecting and customizing a functional system that fits regulatory requirements, specific organizational needs, and those of other stakeholders is one of the most important decisions you will make.

Software Choices

There are many software products available in today's market that significantly reduce the burden of maintaining a manual management information system (MIS). Many of the commercially available systems use some kind of large relational database. For example, one software supplier offers a package that features employee right-to-know software including material safety data sheets for internal use and distribution to customers.[19] Secondary container labeling software is also included, along with industry-accepted hazard ratings. Dolphin also provides a hazard communication training package, chemical approval/purchasing systems, materials management; SARA 312/313, volatile organic compound (VOC), and hazardous air pollutant (HAP) reporting. The system features handling, receiving, and storage procedures; product substitutions; inventory management; waste disposal methods and vendors; report generation; and location tracking.

Another software provider's product offers more than 100 modules to address environmental, health and safety information needs.[20] The modules include tools for project management, documenting general incidents, generating training courses/certifications, producing questionnaires for audits and surveys, and others. Clear the investment of your time with senior management to evaluate off-the-shelf MIS products. Or form an internal implementation team to generate detailed system specifications. The team may employ an MIS consultant to lead the search for a system meeting these specifications. Regardless of the approach, organizations need to avoid "tunnel vision" at all cost. Site-specific and corporate requirements need to drive the search and decision process. Recognize and deal with internal issues like who is to maintain the system, how much information sharing is acceptable, and what degree of support and security is needed. Look for software providers that have a strong, long-term track record and are likely to be available when system changes or updates are needed.

Consider the need for preserving historical information. For example, how do we compare OSHA 200 log injury and illness records with data from the new OSHA 300 recordkeeping system? Note that data conversion packages are needed to make this comparison meaningful.

Above all, the system must be user-friendly. Software designers and MIS people are computer-oriented while users vary considerably with respect to interest and computer-capability. Therefore, systems with menus and clearly presented navigation screens are best. Top management will likely determine the supplier and product you finally select. Use all your influence to assure that your system is designed to handle site-specific MIS challenges and opportunities, especially when time and dollars are in short supply.

Integrate EHS Information Systems

Integrate EHS information systems wherever possible to simplify data entry. Figure 10-7 is a good example of how OSHA and EPA information can be integrated in order to save valuable time entering the data.

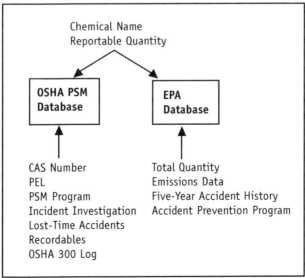

Figure 10-7 Example of Integrating OSHA PSM and EPA Information for One-Time Data Entry[21]

An example of system integration is one that permits one-point data entry from employee and incident information databases, incident reports, and company and state insurance forms. The goal is to design the system so that data is entered once and "the computer generates the calculated answer(s) and places the proper information in all related forms."[22]

You will remember that OSHA's Process Safety Management and EPA's Risk Management Planning regulations list many of the same chemicals. If subject to these regulations, set up the database once, and enter the parameters needed for both OSHA and EPA. For example:

OSHA

- Injury and Illness Information
- Threshold quantity
- Toxicity information
- Permissible exposure limits
- Physical, reactivity and corrosivity data
- Thermal and chemical stability data
- Hazardous effects of inadvertent mixing

EPA

- Threshold quantity (RMP)
- Toxic end point (RMP)
- Emissions data
- On-site storage/usage locations

MIS and Community Outreach

We note from the community-right-to-know discussion above that release and emissions data systems should serve the public's demand for information as well as EPA requirements. Too often, facility or company management does not have the information to answer the public's questions on annual air toxic and carcinogenic emissions, greenhouse gas emissions, energy consumption, process and transportation incidents, and so forth.

Monsanto Environmental, Safety, and Health Project

A few years ago, Monsanto established a project to develop an integrated management information system (MIS). A second objective was to integrate the system with an environmental, safety, and health structure designed to take advantage it. One of the key tasks involved incorporating external and internal changes into their existing operational and compliance processes. The project objectives were to: [23]

- "Transform paper compliance systems to electronic (systems having standard formats) and enable continuous improvement.
- "[Provide an] electronic repository of ESH requirements using standard formats.
- "[Provide] on-line site profile information (like a chemical inventory for each facility).
- "[Allow] electronic storage and sharing of best practices and ESH compliance tools.
- "[Provide] electronic forums for gathering input on proposed ESH requirements. Track issues that are important to (the) enterprise.
- "[Establish] key interfaces (that) include document management systems and external ESH regulatory databases and regulations."

Monsanto used Lotus Notes® as the database-communications network for their system. As a result of the company's experience, the MIS manager related several bits of advice that included "design for and code only the best practices of your business process, involve information technology personnel as early as possible, [and] don't try to put everything in the first release."[24]

EMS Groupware at Anheuser-Busch

Anheuser-Busch took a similar approach in using Lotus Notes® as the communications network. Their system integrated: [25]

1. Environmental, safety, and health information
2. Managerial systems, operational information, and manufacturing information
3. Plant floor systems

Benefits to using Lotus Notes were low cost, ease of creating and modifying simple programs to replace complicated spread sheets, flexibility in modifying the programs, and the ability to blend easily with other company databases.

Enterprise-Wide Solutions

The final information system example is one that allows many users at different locations to communicate, generate, manipulate, exchange, and store common information (Figure 10-8).

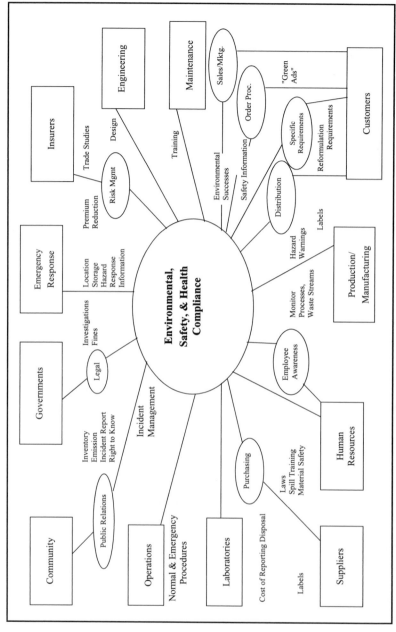

Figure 10-8 Total Organization SHE Information System[26]

Certified safety professional Mark Hansen puts his illustration into perspective. He comments "Enterprise-wide solutions generally allow users throughout the company to communicate, generate, manipulate, exchange, and store common information, which may be generated at several different geographic locations."[27] This type of application requires what is called "fourth-generation program applications software."

Regardless of what you specify to track EHS information, make sure it is integrated into your business MIS system. Think broadly in terms of what elements are needed to drive your overall chemical safety management system. Address internal issues such as control of the system and willingness to share information between locations and system security. Look for applications people and software providers that have a proven track record over the long term.

Implementation Resources

Safety and health resources are found in Chapters 1, 2, and 3. In addition, Table 10-1 lists societies, chemical trade associations, government regulatory organizations, and other sources of information to help the EHS practitioner.

The American Chemistry Council has a wealth of information, most of which is under the banner of Responsible Care®. When using Responsible Care, encourage facilities to do self-reporting early in the process. Be alert to claims of "practice in place" that are not verified. In fact, prevention of premature implementation claims is the main purpose of management systems verification by outside reviewers.

The Center for Chemical Process Safety (CCPS) is an excellent source for management and technical tools that will help to implement an effective chemical safety system. The AIChE/CCPS on-line catalog features over 300 titles—books, magazines, software programs, video tapes, training modules, etc.

AIChE also supports publication affiliates such as the Center for Waste Reduction Technologies, the Design Institute for Emergency Relief Systems (DIERS), the Design Institute for Physical Property Data (DIPPR), and the Process Data Exchange Institute (PDX).

The Chlorine Institute supports the chlor-alkali industry and serves the public through promoting continuous improvement in safety, health, and environmental practices. Their scope of practice includes chlorine, sodium and potassium hydroxides, and sodium hypochlorite, as well as the distribution and use of hydrogen chloride. The institute also distributes pamphlets on design and operation of non-refrigerated liquid chlorine storage facilities, recommendations to chlor-alkali manufacturing facilities for the prevention of chlorine releases, and many others.

The Environmental Defense and Public Interest Research Group consists of advocate organizations. The Environmental Defense "Scorecard" will pinpoint major emission sources given a zip code from a database that comes from the required EPA toxic release inventory reporting (TRI). Environmental Defense lists the top 25 "potentially responsible parties" at the Superfund national priority list (NPL) and features Internet articles critical of Responsible Care.

The Synthetic Organic Chemical Manufacturer's Association directs much of their efforts to the batch and custom chemical industry. Their services include safety and health integrated into chemical process operator training and many other training topics.

Conclusions

Successful implementation of a chemical safety management program takes a balance of planning and action. It is critical that a strategy for continuous improvement of the organization's safety and health infrastructure be part of the business plan.

One of the difficulties encountered here is the tendency for safety efforts to degenerate into "business as usual." For example, the daily pressures of meeting manufacturing targets for chemical A or product B may become a higher priority than the plan to improve use of personal protective equipment. Thus, safety and health must be integrated into business goals and objectives and reviewed regularly to assure that implementation of the management system is on track.

There are 9 elements in the chemical safety management system pictured in Figure 10-1. These are to be installed in a *parallel* rather than in a series of *sequential* steps. Even when new, every organization has strengths and needs relative to these essential elements of safety and health. Use the strengths to address the needs. Constant attention to an annual safety and health plan is a powerful continuous improvement strategy.

Management and safety coordinators seem to know the cause of accidents and what to do about it. It may be obvious, for example, that a spill when transferring sulfuric acid from bulk to tote tanks was caused by failure to follow the proper procedure. Thus, we tell the supervisor to write or re-write the procedure and send the worker off to training. Unfortunately, this apparent insight may do little to prevent the incident from happening again. Do not underestimate the importance of shaping the organization's culture through employee involvement. The culture effect is especially strong in the chemical industry. Involve workers on accident investigation teams and listen to their recommendations. Employee involvement goes a long way to improve safety ownership and control hazards over the long term.

Integrate facility policy with applicable regulatory standards and reinforce constant hazard/risk evaluation and control. Organizational policy must be no more "optional" than complying with an OSHA regulation. Develop an internal and external communication plan. Train and involve all employees to raise the level of safe work practices to habit-strength. Finally, establish information systems that meet facility needs and regulatory requirements.

References

1. "Executive Survey of Workplace Safety," www.libertymutual.com, August 28, 2001.

2. Safety Management Systems (SMS), Ethanol 2000, Bingham Lake, Minnesota, December 2001.

3. Theodore S. Ingalls Jr., "Using Scorecards to Measure Safety Performance, *Professional Safety,* American Society of Safety Engineers, December 1999. Reprinted with permission.

4. Anthony F. Cantarella, "Safety is Something Employees Can Take Home," *Professional Safety,* American Society of Safety Engineers, March 1998.

5. One good source is EnviroWin® Software, Inc., www.envirowin.com.

6. ACGIH®, Inc. www.acgih.org.

7. 29 CFR 1910.119 (j)

8. www.lakeland.com.

9. www.chemrest.com.

10. www.powermation.com.

11. A.M. Best Company, Ambest Road, Oldwick, New Jersey 08858, www.safety.ambest.com.

12. Gelbach Plus Gelbach, "Specializing in community outreach and employee communications," www.gelbach.com and "Effective Community Outreach Planning Workshop," Minneapolis, MN, May 22, 1998.

13. Dr. Fred Millar

14. "Effective Community Outreach Planning Workshop," Minneapolis, MN, May 22, 1998.

15. Occupational Safety & Health Administration, U.S. Department of Labor, www.osha.gov/outreach.html.

16. John M. Beath, "'Right-size' Your Environmental Management Information System," *Chemical Engineering Progress*, American Institute of Chemical Engineers, May 2001.

17. John Dyslin, "Examining the News Recordkeeping Standard," *Safety & Health*, National Safety Council, November 2001.

18. OSHA Form 300, Log of Work-Related Injuries and Illnesses, U.S. Department of Labor, Occupational Safety and Health Administration, www.OSHA.gov/recordkeeping/index.html.

19. Dolphin, Lake Oswego, Oregon 97035, www.dolphinmsds.com

20. DataPipe, Knorr Associates, Inc., Butler, N.J. 07405, www.KnorrAssociates

21. Mark D. Hansen, "Taming the Paper Tiger," *Professional Safety*, American Society of Safety Engineers, October 1998. Reprinted with permission.

22. Hansen, p. 18

23. Michael Cano, "Monsanto IT Project and Groupware System," *Fourth Annual Environmental Management Excellence Program*, September 19, 1997.

24. Cano.

25. Chris Spire, Safety and Environmental Assurance-Anheuser-Busch Inc., *Fourth Annual Environmental Management Excellence Program*, September 19, 1997.

26. Mark D. Hansen, "Taming the Paper Tiger," *Professional Safety*, American Society of Safety Engineers, October 1998. Reprinted with permission.

27. Hansen, p. 20.

Developing a Winning Game Plan

Introduction

The purpose of this chapter is to help the reader develop a winning game plan. We hope we have provided good examples of effective chemical safety management throughout this book. We encourage the chemical safety practitioner to take what is useful. Assess the facility's situation, define problems and opportunities, compare notes, and benchmark other organizations (large and small, established and just starting).

Successful implementation of a chemical safety management system is not about taking a great leap forward. It is about continuous improvement, which depends on careful planning and effective organizing, implementing, and controlling.

Implementation of a chemical management system was discussed in Chapter 10. This chapter will get you started on the steps to success. Safety, health and environmental people are "change agents." Set out to establish good policies and safe work practices. Shape the culture of your organization to improve performance. You can do it!

Make an Assessment and Establish a Driving Force for Change

Chemical safety managers need to help their organization take an inventory of safety and health activities. It is important to link the assessment of safety and health activities to requirements and needs. It is extremely important that the facility understand these requirements and needs to be *criteria* against which they will be measured.

There are at least two ways to assess the current chemical safety management system and install a driving force for change: compliance audits and comprehensive Responsible Care® reviews. The criteria for compliance audits are simply the regulation. The criteria are more broad and flexible for the RC reviews.

Compliance Audits

Process safety audits measure compliance to a fixed regulatory standard. The requirement is to

- Conduct the audits at least every three years and certify compliance
- Choose members of the audit team that are knowledgeable about the regulated process and the audit procedure
- Complete an audit report
- Assure a system for addressing audit findings and documents corrective action follow-up
- Maintain the two most recent compliance audit reports, documenting the corrective actions taken

Note that any documentation that includes recommendations, a report of implementation activities, and evidence of action item tracking/closure technically qualifies as an audit. Facilities should not lead their own PSM audits. Instead, choose an internal team of technical and process safety coordinators from other facilities and led by a trained auditor. The procedure involves document review and interviews with selected managers, as well as with technical, operating, and maintenance personnel. Employ a "teaching style" such that the recommendations focus on continuous improvement rather than simply making yes/no compliance findings.

Large-scale corporate assessments involve multi-disciplined auditors covering elements like: management, process, procedural, medical and health, chemical and physical, plus environmental. Here, the criteria usually combine company/facility and regulatory requirements.

Responsible Care® Codes of Management Practice

Many organizations have formally and informally embraced the principles of Responsible Care®. The program provides a voluntary set of standards, established in 1988 and approved by more than 100 chemical organizations. It is an excellent framework for change and currently used in more than 40 countries throughout the world. The six codes of management practice guide the work of Responsible Care® members and partners:[1]

1. Community Awareness and Emergency Response (CAER)
2. Process Safety
3. Distribution

4. Pollution Prevention
5. Employee Health and Safety (EHS)
6. Product Stewardship (oriented to product-process research and development functions).

Note that the six codes encompass OSHA, EPA, and DOT requirements.

The RC assessment process involves voluntary compliance to criteria established by the American Chemistry Council as "practice in place." The sequence of categories from start to finish is:

NA No action. If no action is taken because the management practice is not applicable, facilities are asked to explain.

EV Evaluation of existing company practices against the RC management practice.

DP Developing plan to implement management practice.

IA Implementing action plan.

PP Management practice in place.

RI Reassessing management practice and implementation.

Follow up, guided by the reviewers' recommendations drive facility activity. Note that there is greater ownership than in the case of regulatory audits. This is because the facility defines activities to meet the objectives. Because they can integrate follow-up to the audits and reviews, the combination makes a significant contribution to regulatory compliance. Note that the reviewers must agree with the facility's definition for practice in place.

The CAER Code

The CAER codes run parallel to EPA's Emergency Planning and Community Right-to-Know requirements in user-friendly language. The portion of Chapter 10 discussing external communications and community outreach provides a summary of how to implement this code.

Process Safety

The process safety code is the voluntary equivalent to OSHA's process safety management regulation (see Table 11-1). It contains 22 statements of management practice designed to help facilities prevent fires, explosions, toxic releases and other process related incidents that has been much of the subject of this book.

Table 11-1 Responsible Care® Process Safety Code of Management Practice and OSHA's PSM[2]

Process Safety Code of Management Practices	1 Commitment	2 Accountability	3 Measurement	4 Incident investigation	5 Information sharing	6 CAER integration	7 Design documentation	8 Process hazards information	9 Process hazards analysis	10 Management of change	11 Siting	12 Codes and standards	13 Safety reviews	14 Maintenance and inspections	15 Multiple safeguards	16 Emergency management	17 Job skills	18 Safe work practices	19 Initial training	20 Employee proficiency	21 Fitness for duty	22 Contractors
OSHA PSM REGULATION 29 CFR 1910.119	X																					
(c) Employee participation		X		X	X	X		X														
(d) Process safety information							X	X			X	X			X			X				
(e) Process hazards analysis									X													
(f) Operating procedures		X					X				X				X	X	X	X				X
(g) Training		X	X					X							X	X	X	X	X	X	X	
(h) Contractors			X									X	X	X			X		X			X
(i) Pre-start safety review		X											X									
(j) Mechanical integrity		X					X					X	X	X			X	X	X	X		X
(k) Hot work permit																	X	X		X		X
(l) Management of change										X												
(m) Incident investigation				X	X																	
(n) Emergency planning and response					X	X					X				X	X				X		X
(o) Compliance audits			X																			
(p) Trade secrets					X	X																
	MANAGEMENT LEADERSHIP						TECHNOLOGY				FACILITIES						PERSONNEL					

Distribution

The distribution code covers management practices in the areas of transportation risk management, compliance review and training, carrier safety, handling and storage, and emergency preparedness. Note that meeting this code fulfills the DOT HAZMAT requirements described in Chapter 8.

Pollution Prevention

The set of pollution prevention codes has to do with the safe management and reduction of wastes. In today's political environment, many companies are reporting annual waste reductions on the internet, even though they may not be Responsible Care® members. Examples of a parallel statutes are EPA's Resource Conservation and Recovery Act (RCRA) and the Comprehensive Environmental Response Compensation and Liability Act (CERCLA) or Superfund Act.

Employee Health and Safety

The EHS code is meant to protect and promote the health and safety of people working at member facilities. It has four general categories: program management, hazard identification and evaluation, prevention and control, and communications and training. Implementing the EHS code will meet the requirements of OSHA's HAZCOM and other standards.

Product Stewardship is a system of designing, manufacturing, distributing, using, recycling, and disposing of products that will minimize negative effects on safety, health, and the environment. EPA's Toxic Substances Control Act (TSCA) is an example of a regulatory requirement related to Product Stewardship.

Organizations are encouraged to take a hard look at RC Guiding Principles and codes of management practice in designing their chemical safety management system. Smaller facilities do not need an extensive set of written management practices such as the RC codes. However, there are significant benefits to using the program's self-evaluation and third party verification system.

Example

An internal PSM auditor may note "In the XYZ process, there is insufficient documentation of process chemistry and the sequence of process flow to determine what process safety information (PSI) was used for the process hazards analysis." He or she will reference (29 CFR 1910.119(d) (2) and (3)). Following an opportunity to discuss the audit finding, the plant will have one to three months to correct the deficiency. Once the audit action item is complete, it is closed through the legal department.

By way of contrast, the RC code is written in more user-friendly chemical safety management language. Reviews of the RC code elements measure and recognize progress made by the plant, not just overall compliance. In the example of PSI used above, several RC code elements come into focus (Figure 11-1): #2–accountability for PSI, #7–design documentation, #8–process hazards information, #11–facility/equipment siting, #12–relevant codes and standards used, and #15–multiple safeguards. The plant therefore has an opportunity to show progress in a number of specific areas that all relate to improvement of process safety information. Because RC reviewers often come from other facilities as well as from staff functions, they bring fresh ideas on how improvement can be accomplished.

Just a few large companies currently using RC principles and what is called *sustainable development and growth* are 3M[3], Oxychem[4], Rohm and Haas[5] and Dupont.[6]

Whether or not your facility uses RC, you will need customized, written EHS programs that are linked to site-specific hazards and reinforce the safe work practices that control them (Figures 10-1 and 10-4).

Integrating Chemical Safety Management Concepts

Chapter 1—Toxic Hazards

The book begins with the introduction of toxicology and industrial hygiene principles. Years ago, it was unusual to find people in plants that were trained in these fields. Today, it is more commonplace, especially in larger chemical industry organizations, to find toxicology and hygiene people in staff positions, in the plants or both.

Toxic hazards are not unique to the chemical and petroleum industries. Many facilities having no chemicals in the process use chlorinated and other solvents for cleaning and degreasing. Most of these materials have regulated exposure limits.

You don't have to be an expert to improve evaluation and control of toxic hazards. Make better use of material safety data sheet information by folding it into facility policies, procedures, and training. Simplify presentation of hazard information (e.g. review the emergency overview section) during hazard communication training. Concentrate on how chemicals enter and affect the body as an introduction to use of personal protective equipment.

Few facilities, especially the small ones, have an adequate medical management system. For example, better monitoring and sampling programs are needed to control air contaminants. Start with a list of job titles and/or exposure situations that qualify for medical surveillance. Employ the latest technology for air monitoring systems (Figure 1-2) to determine actual exposure levels before specifying engineering controls or respiratory protection (PPE).

Apply the safety precedence sequence to the control of toxic hazards. Engineering controls eliminating the hazard should always take priority over safety devices, warnings, PPE, and administrative controls. If engineering controls cannot be applied immediately, identify the hazard and take effective temporary action.

Use sound engineering design principles for process equipment. Contain toxic materials in closed systems. Provide instrumentation and control features that monitor process parameters and include automatic quench systems where appropriate and emergency shutdown.

Finally, contain and recycle potential air, land, and water emissions at the source. Evaluate potential on and off-site impacts from a release. Supplement prevention systems and activities with emergency response planning that involve first responders from the community.

Chapter 2—Fires and Explosions

No chemical safety management system is complete without addressing fires and explosions. These incidents are the leading cause of industrial disasters in U.S. and led to development of OSHA's PSM and EPA's RMP regulations. Fires and explosions can also cause toxic and environmental effects far from the initial source.

Fuel, oxygen, heat, and a chain reaction are the essential components of the fire tetrahedron. The basics of the combustion process provide clues as to how fires start and spread, how they are prevented, and how they are extinguished.

Facility workers, safety coordinators, and emergency responders need a working knowledge of properties of relevant flammable and combustible liquids (Table 2-2). These properties determine important safe work conditions and practices, including storage and handling, transportation and labeling, waste disposal, and emergency response.

Controlling ignition sources is the essential part of fire prevention (Tables 2-1 and 2-6). Electricity, welding/cutting/brazing, hot surfaces and arson are the four ignition sources causing the greatest dollar loss in the U.S. Fire and explosion prevention requirements for flammable and combustible liquids are based on NFPA 30 (Table 2-5) and other NFPA consensus standards. Chemical safety coordinators should identify specific materials and include special circumstances within the facility when planning and implementing fire prevention inspections and hazard studies.

Emergency planning and response is the key element in mitigating the effects of fires and explosions. Often, an effective on-site voluntary emergency squad (Figure 2-2) can prevent a small fire or explosion from becoming a total disaster.

Chapter 3—Design Considerations

Process safety information and other key elements of the engineering design package have great leverage in preventing fires, explosions and chemical releases. Having design standards and safety specification information relating to new and existing facilities is an essential part of accident prevention.

Make sure the design provides multiple independent safety layers (Figure 3-4). Plan design and safety reviews at key stages through the engineering project process.

Assure that significant hazards and risks are identified early in the design stage and that adequate resources are available. Employ inherently safe design principles like "minimize, substitute, moderate and simplify" to eliminate physical and chemical hazards.[8]

Finally, take advantage of technology that provides electronic document management (EDM). Storage and rapid retrieval of design information is an essential tool when analyzing hazards and associated risk.

Chapter 4—Accident and Incident Investigation

Some of the most powerful lessons in life come from our mistakes and safety is no exception. The goal in accident investigation is to prevent similar events from happening again. All accidents, incidents, near accidents and potential hazards should be investigated. Too often, what is perceived as a minor accident is dismissed as "bad luck" or "employee error."

Initiating the emergency action plan is the first step in the accident investigation process. First-aid and accident investigation supplies in the form of a "Go Kit" (Figure 4-2), should be readily available.

Once the scene has been secured and initial alert communicated, the most important decision is to determine the level (degree of real or potential injury/damage and complexity) of the accident. Then the appropriate procedures and participating investigators may be selected. Major accidents and incidents almost always require the services of an outside investigation expert.

Accident investigation policy and procedures need to provide a preplanned process (Figure 4-1). The policy needs to reflect management expectations and values.

Three common mistakes here include:

1. Rapid assignment of blame with no or inadequate investigation
2. The facility safety coordinator is held responsible for accident investigation
3. No initial communication of the incident

Some kind of initial report or accident alert should always precede a full investigation report. Focus the final report on corrective action determined by the safety precedence sequence (Figure 4-6).

Potential process incident investigation procedures come in the form of a process hazards analysis (Figure 4-8). PHA procedures, defined by OSHA's process safety management standard, are also used to assure safe work conditions within a chemical process (Chapter 5).

Chapter 5—Safe Work Conditions

Unsafe work conditions can be a "smoking gun," particularly if there is any evidence that the hazard existed for some time prior to an accident. Most OSHA standards regulate safe work conditions, and the general duty clause (Figure 8-1) covers situations where no specific requirement exists.

Facilities handling chemicals need a particularly strong hazard identification and control program. They must assure safe work conditions and practices associated with all industry, plus those involving acids, bases, flammable liquids, and other hazardous materials.

The chemical safety management infrastructure should include a family of physical department and facility inspections to correct any deviation from the required safe work condition (Figures 5-1 and 5-2). A functioning self-inspection program is essential to correct hazards and build management/employee ownership.

Organizations need documentation such as facility plot plans and building layouts, electrical and process piping and instrumentation diagrams, hazard inspection checklists, plus effective corrective action procedures to assure safe work conditions.

OSHA's PSM standard (Figure 8-4) is the source of regulatory requirements for process-related safe work conditions.

Chapter 6—Safe Work Practices

The initial cause of most accidents involves an unsafe act or failure to follow safe work procedures. The chemical safety manager's attention to written procedures and observation of work practices is the essence of this chapter. Set priorities based on initial and root causes plus contributing factors resulting from accident/incident investigations.

Job safety analysis and the use of behavior methods are the two most common approaches to improving safe work practices. It is important to apply JSA (Figure 6-2) first, especially if hazardous procedures such as chemical sampling and line-breaking are not well documented. Consider some sort of behavior-based system (BBS) for building safe work practices to "habit strength" over the long term.

Chapter 7—Accountability and Performance Measures

EHS coordinators cannot be responsible for all related activities that go on within the organization. A chemical safety management system needs to place accountability with line management, balanced by safety as a condition of employment for all workers.

Integrate EHS and business accountability and objectives. For example, sustainability is increasingly becoming a guiding theme for business decisions. Sustainability means integrating economic growth with environmental stewardship and social responsibility. Embrace sustainable business values that include process, personnel and distribution safety, product stewardship, pollution prevention, community outreach, and compliance to company policies and procedures.

Drive your new game plan via an annual safety and health plan. This tactic helps divide the huge task of continuous improvement into achievable, measurable bites. Top facility management needs to establish personal accountability for safety with visible examples of their commitment to safety (Figure 7-3).

Measure what is important. Result-type measures like recordable and lost-time injuries are outcomes of nonconformance (Figure 7-1) and will always be used to reflect safety performance. Unfortunately, they are of little help in guiding incident prevention activities. Choose upstream or prevention-type performance indicators to help quantify hazard-risk control efforts. Communicate the plan and involve employees at all levels. Acceptance and ownership of the facility performance indicators is especially important to overall program success. Establish and integrate measurements that are indicators of sustainability.

Chapter 8—Regulatory Requirements

Chemical safety coordinators will want to "get their arms around" OSHA, EPA and DOT regulatory requirements as a first step in building a facility management system. It is particularly helpful if the top manager will draft a program reflecting a vision and organizational values about safety (Chapter 7). Once the vision is in place, safety and management functions can to work together to define policy and procedures that meet or exceed regulatory requirements.

Start with OSHA 29 CFR 1910 subparts H and Z as a guideline to customize hazardous materials policy. Then quickly develop policies and procedures addressing the other standards in Table 8-3.

Use one of two methods to assure the policies at least meet the regulations:

1. Apply the regulatory outline to define organizational requirements. Add "comment" or "detail" sections to call out specific instructions related to the facility.
2. Purchase and customize inexpensive software templates to meet regulatory standards in user-friendly language.

An example of the first approach is to outline the elements of the facility process safety management program using the alpha-numeric categories in Figure 8-4. List safety system components under detailed process safety information ((d)(3)(H)). Using the template approach, draft a PSI worksheet, itemizing enclosures, berms, chemical detection systems, rupture disks and pressure relief valves, etc. plus other passive/active components that comprise the safety system. Place the worksheet as an appendix to section 7, Process Safety Information.[9]

Recognize that inspections and audits are driving forces for compliance to OSHA, EPA, and DOT regulations. Do not wait for the agency to visit. Establish teams from selected facilities, led by trained subject-matter auditors. Note that compliance audits are required by the PSM standard every 3 years.

Chapter 9—Safety and Health Training

A brief look at the training program will usually differentiate a good safety and health program from one that is non-existent or inadequate. Effective organizations no longer simply show the video and expect the content to move from the monitor into their employees' memories. Primary benefits from competently trained employees include improved productivity and quality as well as strong safety, health, and environmental stewardship.

Good training takes careful planning. Programs must relate safety and health training to the actual job. Integrating safety into specific jobs and work procedures is an essential component of training (Table 9-4).

Computer-based training is the fastest growing part of the training business. CBT is not the answer to every need (Table 9-5), but can be an important supplement to the basic OJT approach.

Chapter 10—Implementing a Chemical Safety Management System

There was a time when plant managers would fire so-called "accident-prone employees" because they thought that would eliminate future accidents. That approach, often motivated out of frustration, is not needed today by those implementing an effective chemical safety management system (SMS). The indispensable part of a SMS produces results by installing policy based on continuous hazard identification and control. Use applicable regulatory and consensus standards to help define policy and provide detailed procedures that fit the organization. Prioritize hazard identification/control activities and write an annual safety and health plan to accomplish the objectives. Remember that audits and reviews are needed to drive and continuously improve the implementation process.

Facilities often encounter implementation difficulties by not matching policy with activities designed to shape the culture. Employee involvement is critical to ownership and continuous improvement.

Take advantage of what many larger companies have experienced. Remember that they are usually glad to help a smaller organization benchmark their environmental, health, and safety improvement process by means of tours and informational interviews. Seek out other organizations that may have information that will help you. Form consortiums of facilities within an industrial park and integrate community outreach efforts.

A Final Word

The reader may be tempted to set aside an afternoon, boot up the PC, download Table 10-1, print out some items of interest, and put them in a file. While it is good to look up the information of interest, it is better to get more information first-hand. We encourage you to pick up the phone and talk person to person about another practitioner's continuous improvement process and success stories. Make a date to see in person how someone is doing what you want to do. Benchmark companies and facilities that are farther along in the process than you are. Facilities that effectively implement the principles of chemical safety management will be successful. In the end, you will find that safety is as manageable as anything else you do in your organization.

References

1. Responsible Care®, "Good Chemistry at Work," American Chemistry Council, Arlington, Virginia 22209.

2. Adapted with permission from American Chemistry Council (formerly Chemical Manufacturers Association) Responsible Care® Codes of Management Practices. Responsible Care® is a registered service mark of the American Chemistry Council.

3. www.3m.com/about3m/environment.

4. Occidental Chemical Corporation, www.oxychem.com.

5. www.rohmhaas.com.

6. www.dupont.com.

7. Gelbach Plus Gelbach, "Specializing in community outreach and employee communications," www.gelbach.com and "Effective Community Outreach Planning Workshop," Minneapolis, MN, May 22, 1998.

8. T.A. Kletz, *Plant Design for Safety,* Hemisphere Publishing, New York (1991) and Kletz, Taylor & Francis, *Process Plants: A Handbook for Inherently Safety Design*, Bristol, PA (1998).

9. Process Safety Information, p. 10, Process Safety Corporate Safety Plans, Envirowin® Software, LLC. www.envirowin.com.

Using EPA's *Risk Management Program Guidance*

This appendix covers estimating potential toxic release rates for worst-case and alternate scenarios, as well as estimating the distance to the toxic endpoint using EPA's Risk Management Program (RMP) Guidance document and reference tables, RMP*Comp, or commercially available software models. This document can be downloaded from the EPA website referenced below.

The reader may well benefit from a review of Appendix A and perhaps the guidance document itself. Limiting oneself only to RMP*Comp misses important information that may significantly alter the offsite consequence analysis (OCA) from one that may actually occur.

Risk Management Program Guidance for Offsite Consequence Analysis[1]

The purpose of this document (called "*RMP Guidance*") is to help so-called "owners or operators of processes covered by the Chemical Accident Prevention Program rule in the analysis of offsite consequences of accidental releases of substances regulated under section 112(r) of the Clean Air Act."[2] The specific regulation is provided in 40 CFR Part 68. Questions regarding the applicability of this rule to a particular site and other queries may be directed to the EPCRA/CAA Hotline at (800) 424-9346.

Facilities subject to the rule are required to conduct an offsite consequence analysis (OCA) to provide information to the state, local, and federal governments and the public about potential consequences of an accidental chemical release. The OCA consists of two parts:

- A worst-case release scenario
- Alternative release scenarios

The guidelines describe how to predict worse-case and alternative release scenarios by addressing topics including release rates for toxic and flammable chemicals as well as estimation of overpressure distances for flam-

Table A-1 Required Parameters for Modeling (40 CFR 68.22)

WORST CASE	ALTERNATIVE SCENARIO
Endpoints [Para 68.22(a)]	
Endpoints for toxic substances are specified in part 68 Appendix A.	Endpoints for toxic substances are specified in Part 68 Appendix A.
For flammable substances, endpoint is overpressure of 1 pound per square inch (psi) for vapor cloud explosions.	For flammable substance, endpoint is: • Overpressure of 1 psi for vapor cloud explosions, or • Radiant heat level of 5 kilowatts per square meter (kW/m2) for 40 seconds for heat from fires (or equivalent dose), or • Lower flammability limit (LFL) as specified in NFPA documents or other generally recognized sources.
Wind speed/stability [Para 68.22(b)]	
This guidance assumes 1.5 meters per second and F stability. For other models, use wind speed of 1.5 meters per second and F stability class unless you can demonstrate that local meteorological data applicable to the site show a higher minimum wind speed or less stable atmosphere at all times during the previous three years. If you can so demonstrate, these minimums may be used for site-specific modeling.	This guidance assumes wind speed of 3 meters per second and D stability. For other models, you must use typical meteorological conditions for your site.
Ambient temperature/humidity [Para 68.22 (c)]	
This guidance assumes 25 deg. C (77 deg. F) and 50 percent humidity. For other models for toxic substances, you must use the highest daily maximum temperature and average humidity for the site during the past three years.	This guidance assumes 25 deg. C and 50 percent humidity. For other models, you may use average temperature/humidity data gathered at the site or at a local meteorological station.
Height of release [Para 68.22 (d)]	
For toxic substances, you must assume a ground level release.	This guidance assumes a ground-level release. For other models, the release scenario may determine release height.
Surface roughness [Para 68.22 (e)]	
Use urban (obstructed terrain) or rural (flat terrain) topography, as appropriate.	Use urban (obstructed terrain) or rural (flat terrain) topography, as appropriate.
Dense or neutrally buoyant gases [Para 68.22 (f)]	
Tables or models used for dispersion of regulated toxic substances must appropriately account for gas density. If you use this guidance, see Tables 1-4 for neutrally buoyant gases and Tables 5-8 for dense gases, or Tables 9-12 for specific chemicals.	Tables or models used for dispersion of regulated toxic substances must appropriately account for gas density. If you use this guidance, use Tables 14-17 for neutrally buoyant gases and Tables 18-21 for dense gases, or Tables 22-25 for specific chemicals.
Temperature of released substance [Para 68.22 (g)]	
You must consider liquids (other than gases liquefied by refrigeration) to be released at the highest daily maximum temperature, from data for the previous three years, or at process temperature, whichever is higher. Assume gases liquefied by refrigeration at atmospheric pressure to be released at their boiling points. The guidance provides factors for estimation of release rates at 25 deg. C or the boiling point of the released substance, and also provides temperature correction factors.	Substances may be considered to be released at a process or ambient temperature that is appropriate for the scenario. This guidance provides factors for estimation of release rates at 25 deg. C or the boiling point of the released substance, and also provides temperature correction factors.

mable chemicals. Reference or "look-up tables" are provided for all covered substances and conditions required for reporting. Six appendices include references for consequence analysis methods, data for toxic substances and mixtures, flammable substance information, technical background, OCA worksheets, and a copy of the RMP regulation.

Table A-1 is a reproduction of the release and dispersion parameters required for modeling worst-case and alternative scenarios.

The result of the analysis is the distance to an offsite endpoint for regulated and potentially released toxic and flammable substances. The RMP Guidance document contains 12 chapters of information to help the user develop a hazard assessment that meets EPA requirements. The information is oriented to application but relates strongly to the theory of modeling.

Note that EPA has developed industry-specific guidelines to help users define their RMP for:

- Propane storage facilities
- Chemical distributors
- Waste water treatment plants
- Warehouses
- Ammonia refrigeration
- Small propane retailers & users

The guidance document provides a step-by-step procedure for determining the offsite impact of a toxic or flammable release. The steps for a toxic gas or liquid are:

1. Define the scenario.
2. Select the scenario.
3. Calculate the release rates.
4. Find the toxic endpoint.
5. Determine the appropriate reference table and distance to endpoint.

Step 1—Define the worst case or alternative scenario

EPA defines a worst-case release as the release of the largest quantity of a regulated substance (toxic or flammable) from a single vessel or process line failure that results in the greatest distance to an endpoint. The distance to the endpoint is the distance a toxic gas/vapor cloud, heat from a fire, or blast waves from an explosion will travel in order to "present imminent and substantial endangerment to public health and the environment."[3] Another way to say this is that at the endpoint, the toxic cloud or heat/blast wave has dissipated to the point that serious injuries are unlikely to occur.

Possible causes or the probability of a worst-case release are not factors to be considered; the worst-case release is assumed to take place. This requirement is the most controversial part of the standard because of anticipated adverse public reaction to the worst-case scenarios. EPA admits that the assumptions required are intended to provide results that overestimate the effects of a release.

Worst-case scenarios for toxic gases must assume that the entire quantity of material is released in 10 minutes. Passive mitigation factors (such as release inside a building or other enclosure) may be taken into effect.

For toxic liquids, the total worst-case quantity is assumed to spill onto a flat, non-absorbing surface. If the liquid is carried in pipelines, you must assume the release forms a pool from which it evaporates to the atmosphere.

Worst-case flammable releases are assumed to result in a vapor cloud containing the total quantity of material.

Quantity of released gas or liquid must be the largest amount contained in a single vessel or pipe. Administrative controls that limit the quantity may be taken into account.

Alternative release scenarios are releases that are more likely to occur than the worst-case situation. Again, this offsite consequence analysis (OCA) is to be conducted to reach an endpoint offsite unless no such scenario exists. Facilities have flexibility to choose alternative release scenarios appropriate to their site(s). Examples are:

- Releases due to splits in transfer hoses or sudden hose uncoupling
- Releases from process piping due to failure at flanges, joints, welds, valves/seals, and drain or bleed lines
- Overfilling vessels to the point of spill
- Release through rupture disks or relief valves due to over-pressurizing vessel(s)
- Spill due to puncturing or mishandling shipping containers

Alternate release scenarios for toxic or flammable materials should be those that lead to an impact beyond the fence line. Situations such as startup, shutdown, and other non-routine conditions may be considered.

Passive and active mitigation measures that are in place such as interlocks, automatic shutdown systems, pressure relief, and deluge systems may be considered to reduce the impact of an alternative scenario.

Step 2—Select the scenario

In selecting the scenario, the reader should first examine the regulation carefully (subpart B, 40 CFR 68.25 for worst-case and 40 CFR 68.28 for alternative release scenarios).

Several factors need to be taken into consideration. Only the hazardous property (toxic or flammable) for which the material is regulated must be considered. For example, ammonia is toxic and flammable, but only toxicity must be considered in selecting the scenario. If a regulated flammable material is also toxic, only flammability is considered.

Facilities that only have program 1 processes (lowest requirement) must report the process and worst-case scenario causing the greatest distance to an endpoint. For program two or three processes, facilities must report on one worst-case analysis representing all toxic regulated substances above the threshold quantity and one worst-case analysis for all flammable regulated materials above the threshold. For more than one regulated substance in a class, choose the one that has the potential to cause the greatest offsite impact.

There are more choices allowed in selecting alternate release scenarios.

* Facilities may choose the worst-case scenario and apply active mitigation measures in place to reduce the quantity and duration of the release.
* The scenario may be selected from a process hazards analysis conducted by the facility.
* An actual event may be chosen by reviewing the accident history for a regulated process.
* Means other than a process hazards analysis may be used to determine possible system failures and release scenarios.

The key information needed in selecting the alternate scenario is quantity released and the time-duration of the release. This information will allow calculation of the release rate and proceed in the same manner as for the worst-case scenario.

Step 3—Calculate the release rates

The guidelines give specific instructions for:

- Toxic gases
  - ❑ Unmitigated
  - ❑ Passive mitigation
  - ❑ Refrigerated
- Toxic liquids
  - ❑ Releases from pipes
  - ❑ Unmitigated
  - ❑ Passive mitigation
  - ❑ Releases of mixtures
  - ❑ Release temperature corrections

Estimating Release Rates for Toxic Gases

Regulated materials that are gases at ambient temperatures (25 degrees C, 77 degrees F) are considered gases for the consequence analysis unless the gas is liquefied by refrigeration at atmospheric pressure. Gases liquefied by pressure alone are treated as gases.

Gases that are liquefied by refrigeration and released into diked areas are treated as liquids at their boiling points if they would form a pool that is more than one centimeter (0.3937 inches) in depth. If the pool depth is one centimeter or less, the material should be treated as a gas.

Note that modeling of these refrigerated systems shows that the evaporation rate from such a pool would be equal to or greater than the dispersion rate for a toxic gas released over 10 minutes. Therefore, treating liquefied refrigerated gases as gases rather than liquids in such cases is a reasonable assumption.

Passive mitigation (e.g. dikes or enclosures) may be considered for worst-case gaseous releases and releases of gases liquefied by refrigeration.

Passive Mitigation

The best example of passive mitigation is a release of a toxic gas in an enclosed space. Modeling a gaseous release inside a building, shed, or other enclosure lessens the release rate to the outside air considerably. The factors involved in this type of release are quite complex. However, the facility person is advised to use a simplified method to approximate release to the outside air from a release rate into an enclosed space. The method assumes that the release occurs in a fully enclosed, non-airtight space that is directly adjacent to the outside air. When modeling a release in an interior room that has no exterior walls, a greater mitigation factor may be more appropriate. Similarly, less mitigation should be used for a space with doors or windows that could be open during a release. EPA advises facilities to perform *site-specific modeling* if any of these special considerations apply. In addition, the passive mitigation effect should not be used if there is reason to believe the structure would not survive the force of the release or if the chemical is handled outside the building (e.g., moved from one building to another).

Gaseous Releases Inside Buildings

EPA has based the mitigation factor for gaseous release inside a building on a document entitled *Risk Mitigation in Land Use Planning: Indoor Releases of Toxic Gases,* by S.R. Porter.[4] The study considered release rates of up to 2,000 pounds per minute, roughly the maximum pressure buildup that most buildings could withstand. Data is provided for the maximum release rate inside a building and the corresponding maximum release rate from the building.

Using lower (3.36 kg/s) and upper (10.9 kg/s) maximum release rates, and allowing for the time for the release to accumulate in the building, Porter determined that the release rate from a 1,000 cubic meter building is roughly 55 percent of the rate into the building.

Porter's paper presented three different ventilation rates, 0.5 (representative of specially designed "gas-tight" buildings), 3, and 10 air changes per hour. EPA determined that 0.5 was appropriate for the entire analysis. Thus a mitigation factor of 55 percent may be used in the event of a gaseous release which does not destroy the building into which it is released. This factor may overstate the mitigation provided by a building with a higher ventilation rate.

For the worst case, we must assume the largest quantity resulting from a pipe or vessel failure is released over a 10-minute period. Thus, the mitigated release rate is estimated as follows:[5]

$$QR = \frac{QS}{10} \times 0.55$$

It is not unusual for facilities to build a simple structure around a gaseous storage tank or a process in order to take advantage of the 55% mitigation factor. Several 3M facilities have a *penthouse* to contain toxic gases that might otherwise be released directly to the atmosphere from blow down tanks above the process.

Releases of Liquefied Refrigerated Toxic Gas in Diked Areas

For situations where the toxic gas is liquefied by refrigeration only, the regulation specifies that the worst-case scenario involves evaporation from a liquid pool at the boiling point of the liquid. The calculation depends on whether the depth of the liquid is more or less than 1 centimeter (0.3937 inches). If less than 1 centimeter, treat the release like an unmitigated gas.

For release in a diked area, we must first compare the diked area to the maximum area of the pool that could be formed. The maximum size of the pool may be estimated by the equation:[6]

$$A = QS \times DF$$

where: A = Maximum area of pool (square feet) for depth of 1 cm

        QS = Quantity released (pounds)

        DF = Density Factor for toxic gases at their boiling points[7]

If the pool formed by the released liquid would be smaller than the diked area, assume a 10-minute gaseous release and estimate the release rate as unmitigated. If the dikes prevent the liquid from spreading out to form a pool of maximum size (one centimeter in depth), you may use the method for mitigated liquid releases but at the boiling point of the released liquefied gas:[8]

$$QR = 1.4 \times LFB \times A$$

where: QR = Release rate (pounds per minute)

        1.4 = Wind speed factor = $1.5^{0.78}$,

             where 1.5 meters per second (3.4 miles per hour) is the wind speed for the worst case

        LFB = Liquid Factor Boiling[9] (for liquefied gases)

        A = Diked area (square feet)

After estimation of the release rate, estimate the duration of the vapor release from the pool (the time it will take for the pool to evaporate completely) by dividing the total quantity spilled by the release rate. Knowing the duration of the release will allow you to choose the appropriate reference table of distances from the RMP Guidance document and estimate the consequence distance. (You do not need to consider the duration of the release for chlorine or sulfur dioxide, liquefied by refrigeration alone. Only one reference table of distances is provided for worst-case releases of each of these substances, and these tables may be used regardless of the release duration. The principal reason for making no distinction between 10-minute and longer releases for the chemical-specific tables is that the differences between the two releases are small relative to the uncertainties that have been identified).

Chapters 6 and 7 of the *RMP Guidance* document discuss release rates for alternative scenarios. Section 7.1 of the document provides a discussion of release rates for toxic gases, and section 7.2 covers the same for toxic liquids. Examples of both unmitigated (no passive or active means of protection) and mitigated release situations are provided for diborane, hydrogen fluoride, allyl alcohol, and bromine.

Release of Toxic Liquids

EPA's release rate to the atmosphere for toxic liquids is assumed to be the rate of evaporation from the liquid pool. For the worst-case analysis, we assume the total quantity in a vessel or the maximum quantity coming from pipes is released into the pool. As in the case of refrigerated liquids, the practitioner may consider passive mitigation measures like dikes to determine the area of the pool, the release rate, and how long it takes for the entire pool to evaporate (duration of release). The RMP rule requires the assumption that liquids are released at the highest maximum daily temperature for the previous three years or at process temperature, whichever is higher. The guidelines provide methods to estimate the release rate at 25 ºC (77 ºF), at the boiling point and a method to correct for release temperatures between 25 ºC and 50 ºC. The calculation methods apply to substances that are liquids under ambient conditions and that are released to form pools deeper than one centimeter.

Release of Toxic Liquids From Pipes

Consideration of a liquid release from a broken pipe involves the maximum quantity that could be released. The EPA guidelines (section 3.2.1 of the *RMP Guidance* document) perform the calculation based on the assumption that the entire pipe is full of liquid. Thus the volume of liquid released is the full length of pipe times the cross-sectional area. The mass quantity is then calculated from the volume divided by a specific density factor[10] for the liquid chemical. Assume the mass quantity (in pounds) is released into the pool. Use methods for unmitigated and for passive mitigation to calculate the evaporation rate of the liquid from the pool.

Other Subjects in the Guidance Document for Toxic Liquid

Endpoints and other data for toxic gases are found in the appendix, exhibit B-1 of the *RMP Guidance* document. Users can find similar data for toxic liquids (exhibit B-2) and water solutions (exhibit B-3).

Exhibits B-1 and 2 will indicate if the gas/vapor cloud is neutrally buoyant or dense. For worst case scenarios, use reference tables 1-12 to look up the end point distance, depending on whether the gas/vapor density is neutrally buoyant, dense, or one of the three chemical-specific cases. The distance to the endpoint is either given from a table as a function of the release rate divided by the toxic endpoint or from a table as a function of release rate and the toxic end point.

For alternate scenarios, use corresponding reference tables 14-25. The tables are similar to those for finding worst-case distances.

Note that a toxic gas such as ammonia that is normally lighter than air may act as a dense gas when released if it is liquefied under pressure (because the released gas may be mixed with liquid droplets) or if it is cold. Thus the state of the released material must be considered before selecting the reference table.

Example- Ammonia

Consider the worst-case, unmitigated release of 20,000 of anhydrous ammonia, either non-liquefied or liquefied by refrigeration. Exhibit B-1 tells us that the toxic endpoint is 0.14 mg/L and that the gas/vapor density is buoyant. Then the proper reference (table 10) indicates the distance at a release rate of 2000 lbs./min. to be 2.2 miles (rural) and 0.8 miles (urban).

If the ammonia is liquefied under pressure, it will release as a dense gas. Then reference table 9 indicates the distance to be 2.6 miles (rural) and 1.7 (urban).

References

1. The techniques and steps used to estimate and analyze scenarios in this Appendix have been adapted from the EPA document *Risk Management Program Guidance for Offsite Consequence Analysis*, United States Environmental Protection Agency, Office of Solid Waste and Emergency Response (5104), EPA 550-B-99-009, April 1999, www.epa.gov/ceppo.

2. *RMP Guidance*, introduction.

3. 40 CFR 68.3, Appendix 1, Definitions.

4. *RMP Guidance*, Appendix D, p. D-1.

5. *RMP Guidance*, Equation 3-2, p. 3-3.

6. *RMP Guidance*, Equation 3-6, p. 3-8.

7. Density factors are listed in the *RMP Guidance* document, Exhibit B-1, Appendix B.

8. *RMP Guidance*, Equation 3-8, p. 3-8.

9. Liquid boiling factors are listed in the *RMP Guidance* document, Exhibit B-1, Appendix B.

10. Density factors are listed in *RMP Guidance* document, Exhibit B-2, Appendix B.

Modeling Software

This appendix is the companion to Appendix A and describes the use of modeling software in order to meet EPA's requirement for offsite consequence analysis (OCA) of a potential toxic release, fire, or explosion. Choices for release and dispersion software include EPA's RMP*Comp, ALOHA®, and SAFER SYSTEMS®.

Note that these offsite studies have been fraught with legal battles, attempts to override EPA guidelines, demonstrations that standard EPA input assumptions are too conservative for specific site use, etc. The reader is well advised to seek assistance of a dispersion meteorologist. Further, such critical analyses need good on-site meteorological data to be used as input to the standard models. Generic data often yields inaccurate results. For example, data taken from the nearest weather station (NWS) may be accurate but not representative of the site conditions. Average wind direction may be skewed and/or average vertical thermal data may be different from NWS data. Managing and validating a five-year database is detailed and time-consuming work. This work should be done by a meteorologist who knows how data are used by a dispersion model and what kinds of missing data clumps might skew the outcome. Professional meteorologists speak of frequent failures of standard data to accurately represent site conditions, making important studies invalid.

RMP*Comp

RMP*Comp[1] is the free EPA software version of the EPA *RMP Guidance* document (Appendix A), although the two models do not always provide equivalent estimates. RMP*Comp was developed by the CAMEO (Computer Aided Management of Emergency Operations) Team at the Hazardous Materials Response Assessment Division, NOAA (National Oceanic and Atmospheric Administration Office of Response and Restoration) and the Chemical Emergency Prevention and Preparedness (CEPPO) section of the EPA.

Table B-1 Endpoints for Selected Toxic Gases and Liquids Using RMP*Comp Version 1.06

Chemical	Liquid, Gas, and Gas/Vapor Density	Release Rate (lbs/min), Duration (min.)	EPA Threshold Quantity (TQ), 40 CFR Part 68 (lbs.)	Toxic End Point, mg. per liter	Worst Case Distance to Endpoint (miles) for TQ (1)		Alternative Scenario Distance to Endpoint (miles) for ½ TQ (2)	
					Rural	Urban	Rural	Urban
Acrylonitrile	L, dense	307 lbs/min., 65 min.	20,000	0.076	4.1	2.9		
Anhydrous Ammonia	G, dense	1000, 10	10,000	0.14	1.8	1.2	0.8	0.3
Chlorine	G, dense	250, 10	2,500	0.0087	3.4	1.5	0.3	0.1
Ethylene Oxide	G, dense	1000, 10	10,000	0.090	3.6	2.5		
Hydrogen Fluoride	G, buoyant (may be dense)	100, 10	1,000	0.016	3.0	1.6		
Methyl Hydrazine	L, dense	6465, 2	15,000	0.0094	7.4	5.7		
Phosgene	G, dense	50, 10	500	0.00081	11	8.1		
Sulfur Dioxide	G, dense	500, 10	5,000	0.00078	5.2	2.1	0.3	0.1
Toluene Diisocyanate	L, buoyant	0.0336, 4960 hrs.	10,000	0.007	0.1	0.1		
Vinyl Acetate Monomer	L, dense	289, 52	15,000	0.26	1.9	1.2		

(1) Unmitigated, ammonia liquefied under pressure, dense gas/vapor cloud.
(2) 1" diameter hole in tank, height of liquids above hole = 24", tank pressure for gases = 2 atmospheres absolute pressure, chemical temperature = 77 degrees F, estimate release rate, no mitigation.

RMP*Comp users must provide parameters for the release according to the RMP standard and the Guidance document. This includes quantity released, release rate and duration, mitigation measures, and topography (either rural or urban). Buildings and trees increase mixing of toxic releases with the air, while lakes and open areas provide decreased mixing. Thus, rural topography predicts longer distances to the end point.

Table B-1 presents examples of worst-case and alternative distances to endpoints for selected toxic gases and liquids, using version 1.06.

Compared to the Guidance document lookup tables, RMP*Comp has significant advantages. RMP*Comp saves time and hand calculations are eliminated, reducing the possibility for error. Both systems result in very conservative endpoint estimates for toxic chemicals. That is, predicted "footprint" distances are greater than those resulting from actual release and dispersion conditions.

For flammable materials, RMP*Comp results are not very sensitive to variations in the parameters from the table. Worst case scenarios at the threshold quantity (10,000 lb) for butadiene, acetaldehyde, dimethyl amine, and hydrogen show little difference to the 1 psi overpressure endpoint. Distances for these four flammables (in a gaseous state, not liquefied) fall within 0.2 to 0.3 miles.

RMP*Comp determines flammable material endpoints for alternate scenarios resulting in vapor cloud explosions, vapor cloud fires and BLEVEs (boiling liquid, expanding vapor explosions). Little difference was found between distances and chemicals for release rates in the 1000–5000 lb per minute range. Most of the alternative scenario endpoints calculated for these four flammable materials fell into the range of less than 0.1–0.2 miles.

RMP*Comp is free and convenient, but remember that you get what you pay for. Be sure to download the latest version from the EPA/CEPPO website.

ALOHA®

ALOHA®, available to the public and developed by EPA and NOAA, is part of the CAMEO® package. CAMEO® consists of 12 information modules, 2 major applications (MARPLOT and ALOHA®) and an application, SPV, to view facility diagrams.

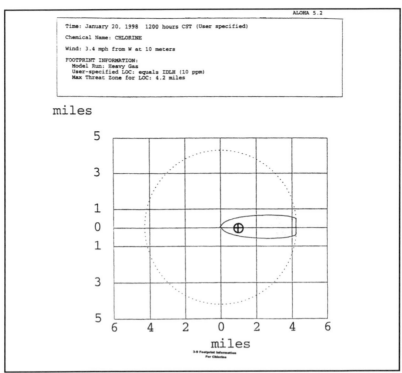

ALOHA 5.2

Time: January 20, 1998 1200 hours CST (User specified)

Chemical Name: CHLORINE

Wind: 3.4 mph from W at 10 meters

FOOTPRINT INFORMATION:
  Model Run: Heavy Gas
  User-specified LOC: equals IDLH (10 ppm)
  Max Threat Zone for LOC: 4.2 miles

Figure B-1 Footprint Information for Chlorine Using ALOHA®

ALOHA® is a good model for obtaining worst-case and alternative distances to report in the RMP hazard assessment. It is also useful for emergency planning and response because distances and dose values can be determined and modified quickly to match conditions at the site. Note that it is not suitable for fire or overpressure calculations. It may be downloaded from the EPA website.[2]

Figure B-1 illustrates the ALOHA footprint for a worst-case unmitigated chlorine release of 20,000 lbs.[3]

ALOHA® predicts a maximum threat zone of 4.2 miles for this release. Note that the cloud shape is no longer a circle, but more accurately accounts for direction and magnitude of the wind. This model has also been used to determine the chlorine concentration 1.0 miles downwind where a hypothetical school exists. The concentration outside the school would be 382 parts per million (ppm) and inside the school equal to 20.7 ppm or 0.06 mg/liter. The OSHA permissible exposure limit for chlorine is 0.003 mg/liter (8 hour ceiling time-weighted average). Adverse health effects could result for students and teachers if this exposure actually occurred.

The ALOHA model has an artificial distance cutoff of 6 miles. Users should consider choosing a different model if the scenario would likely result in a greater distance.

ALOHA may also be downloaded from the EPA/CEPPO website. There is a two-level charge for the software—one price for government and nonprofit organizations and another for commercial applications.

SAFER SYSTEMS®[4]

Figure B-2 displays a representative vapor cloud profile using the SAFER/ TRACE system.

This proprietary, modular software provides comprehensive capabilities including calculation of release rates, dispersion modeling, weather data analysis, and consideration of complex terrain.

The hazard assessment feature contains evaporation and atmospheric dispersion modeling tools for a full-range of release types. Snapshots and vapor cloud footprints are displayed as well as impacted receptors. Chemical specific toxicity reports can be generated. The module can interface to real-time meteorology and sensor data acquisition modules.

The release rate estimation enables users to calculate the release rate from a pipe leak, tank rupture, or the failure of a pipe attached to a tank.

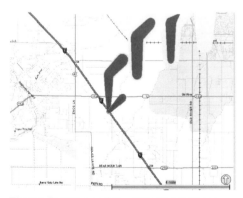

Figure B-2 Vapor Cloud Footprint Using SAFER Systems®[5]

The model can incorporate site-specific topographical features. Plume dispersion profiles conform to the terrain and are consistent with meteorological measurements.

Infiltration and ex-filtration features calculate the indoor concentration profile given air exchange information and outdoor concentrations. The ex-filtration module calculates the release rate from a source within a structure and determines the dispersion profile through the atmosphere.

The SAFER system accounts for regions of complex terrain where multiple weather stations may be required to account for changes in wind speed and direction.

The software includes a customization tool-kit and utilities so that modifications may be made to site-specific maps, menus, and reports. Hazard levels may be modified based on chemical properties.

Special algorithms may be used to replace standard methodologies in special situations that include enhanced evaporation models and atmospheric dispersion of chemicals involving specific reaction dynamics.

Release Rates may be back calculated by using concentration measurements from multiple locations and site-specific meteorological data. It is possible to back-calculate the release rate and compare predicted concentrations to actual measured values.

Instantaneous data may be collected in real time from one or more on-site meteorological stations and used to calculate five-minute averaged values. The values are stored and may be used for modeling and for updating projections of impacted areas when weather condition changes occur.

Stored data can be retrieved and statistical analysis performed. Frequency graphs may be generated to show variable directional effects and the data can be exported to external programs.

The program can quickly display either a "toxic corridor" or a "fugitive emissions corridor" to help locate an impact sector or release source. Meteorological data of the station selected is displayed and a report may be generated listing receptors lying within the toxic corridor.

Concentration measurements are obtained from gas detection sensors and can be displayed and stored. Alarm messages are activated and warnings issued at preset levels. There is a network setup to transmit and view information at other computers.

Comparison of Models

EPA clearly states that results using the reference tables or RMP*Comp are expected to be conservative and will generally, but not always, overestimate the distance to endpoints. The chemical-specific reference tables are less conservative than their generic counterparts. EPA admits that more complex models may give significantly less conservative estimates but point out that they may be expensive and require more expertise to use. Users need to consider their specific requirements and the trade-offs involved in selecting a model. The point is that modeling results will often vary considerably from model to model depending on what parameters are used and the assumptions made.

An independent comparison of the EPA guideline tables with a commercial model (PHAST/SAFETI Version 5.2) shows that the guidelines oversimplify release source modeling.[6] EPA guidelines use a "specific model (SACRUNCH) for ammonia releases and a generic guideline (SLAB) for all other materials." The authors describe differences obtained using a variety of gases (ammonia, chlorine, sulfur dioxide) and liquids (acrylonitrile, carbon disulfide). Other variables that produced differences from the guidelines included release direction and "the actual phase of a release, whether from the liquid or vapor space."[7]

Studies like this one and comparison of other models with the EPA guidelines should influence users to examine site chemicals, quantities and potential release conditions carefully before selecting a model. If RMP*Comp predicts an offsite impact, it is a good idea to consider using other models.

Figure B-11 summarizes the characteristics of several common models including the EPA lookup tables. For comparison, consider RMP*Comp equivalent to the lookup tables.

It is important to recognize that hazard distances often depend on variables not considered in EPA guidelines or part of RMP*Comp. Factors such as release direction (up, down, and horizontal), actual phase of a release (whether from the liquid or vapor space), and release height can have a big impact on the results. More sophisticated software such as SAFER Systems and the proprietary model PHAST take these variables into account and show significant differences compared to the EPA guidelines. Facilities need to carefully examine facility/process siting, potential release parameters, and location of potential impact zones before choosing a dispersion model.

Table B-2 Comparison of Models For Worst-Case Scenario Evaluation[8]

Model	Public/Proprietary	Model Capability	Strengths	Limitations	Comments
EPA Offsite Consequence Analysis (OCA) Guidance (Lookup Tables)	Public; developed by EPA	Gas or liquid Buoyant or dense modeling Mixtures Explosion overpressure	Don't have to justify EPA published Openly available Credit for some passive mitigation (e.g. dikes)	Gives very conservative results for toxics Hand calculations required Prone to data errors Limited flexibility	Provides first-pass prediction Models up to 25 miles Commercial software is available Remains in draft status
ALOHA	Public; developed by EPA and NOAA	Gas or liquid Buoyant or dense gas modeling	Used by many local agencies (<3,000 copies) Models hundreds of chemicals Easy to use Less conservative than OCA	Does not do fire or explosion overpressure calculations	Truncates (cuts short) at 6 miles Does not automatically calculate gas emissions from an indoor release
SAFER/TRACE	Proprietary; developed and supported by SAFER Systems	Gas or liquid Buoyant or dense gas modeling Explosion overpressure	Preloaded chemicals Does release rate calculations Has graphics and map capability	Not readily available to emergency agencies Limited distribution in industry Model basis is proprietary and must be communicated	SAFER truncates at 2 hours TRACE provides dispersion >25 miles

Table B-2 Comparison of Models For Worst-Case Scenario Evaluation (continued)[8]

Model	Public/ Proprietary	Model Capability	Strengths	Limitations	Comments
DEGADIS	Public; co-funded by EPA, DOE, and DOT	Gas or liquid Dense gas modeling	Windows-based Easy to use Chemicals can be preloaded Portions of model are incorporated into ALOHA	Model basis must be communicated Needs expert support Limited chemical database can be supplemented	Provides dispersion >25 miles
PHAST	Proprietary; developed by Det Norske Veritas	Gas or liquid Buoyant or dense gas modeling Can handle ideal explosion overpressure	DIPPR chemical database Can handle aerosols Previous releases widely accepted within industry Good graphics and map capability	Dispersion may exceed OCA Not readily available to emergency agencies but probably known to LEPCs Model basis is proprietary and must be communicated	Used to model many of Kanawah Valley scenarios Provides dispersion>25 miles
HGSYSTEM	Public; developed by joint industry/agency group	Gas or liquid Buoyant or dense gas modeling Can handle mixtures	Very good for ammonia and hydrogen fluoride and some other dense gases Joint industry/government validation	Very difficult to use; expert support needed Does not do overpressure Prone to input mistakes No graphics capability	Complicated to use; model output is confusing Provides dispersion >25 miles

References

1. RMP*Comp software can be downloaded from the EPA/CEPPO Internet website at www.epa.gov/ceppo.

2. www.epa.gov/ceppo/cameo/request.htm.

3. ALOHA 5.2, Case 3, "Dispersion Modeling, EPA Risk Management Program (RMP): Accidental Release Prevention Requirements," University of St. Thomas, Minneapolis, MN, March 6, 1998.

4. Available from SAFER Systems, L.L.C., 4265 E. Thousand Oaks Blvd., Suite 350 Westlake Village, California, www.safersystem.com.

5. Reprinted with permission from SAFER Systems, L.L.C.

6. John L. Woodward, David R.E. Worthington, "Comparison of EPA Guidelines Tables With a Commercial Model," *Process Safety Progress,* Vol. 18, No. 1, Spring 1999, Figure 1.

7. Woodward, Worthington, p. 25.

8. Kevin J. Mitchell and Jatin N. Shah, "Selecting Hazard Assessment and Communications Techniques," *Chemical Engineering Progress,* September 1998. Reproduced with permission of the American Institute of Chemical Engineers. Copyright 1998. All rights reserved.

Glossary

ACGIH–American Conference of Governmental Industrial Hygienists, a group of government industrial hygienists that sets and reviews safety and health standards.

Acute exposure–an almost instantaneous exposure in which rapid changes occur. An acute exposure is shorter and the effects are easier to reverse compared to a chronic exposure.

Auto ignition temperature–see ignition temperature.

Boiling liquid expanding vapor explosion (BLEVE)–A BLEVE occurs if a vessel ruptures which contains a liquid at a temperature above its atmospheric-pressure boiling point. The subsequent BLEVE is the explosive vaporization of a large fraction of the vessel contents; possibly followed by combustion or explosion of the vaporized cloud if it is combustible. This type of explosion occurs when an external fire heats the contents of a tank of volatile material. As the tank contents heat, the vapor pressure of the liquid within the tank increases and the tank's structural integrity is reduced due to the heating. If the tank ruptures, the hot liquid volatilizes explosively.

Catastrophic release–a major uncontrolled emission, fire, or explosion, involving one or more highly hazardous chemicals, that presents serious danger to employees and offsite receptors.

Chemical–any element, compound, or mixture of elements and/or compounds.

Chronic Exposure–illnesses developed over a relatively long exposure period, often having a long-lasting effect. An example of the effect of a chronic exposure is long-term irritation of the nasal passage due to repeated inhalation of hazardous vapors.

Consensus Standards–national standards that represent general agreement among representatives of various interested or affected organizations and individuals. Consensus standards state very specific safety requirements and are considered "teaching standards."

Deflagration–an explosion where the shock wave, reaction or flame front moves at less than the speed of sound in the un-reacted material.

Detonation–an extremely rapid reaction usually resulting in an explosion. The shock wave, reaction or flame front moves at greater than the speed of sound in the un-reacted material.

Engineering controls–a physical method of eliminating or controlling a hazard. Methods include inherent safe process and equipment design, use of isolation and enclosures, and ventilation.

Failure modes and effects analysis (FMEA)–a hazard evaluation technique used to evaluate possible equipment failure modes and evaluates the consequences on the system. Equipment failure modes may be either initiating or contributing events leading to an accident. This method is ideal for single equipment failures and does not readily lend itself to combinations of failures.

Fault Tree Analysis–a hazard evaluation technique that diagrams (in the form of a tree) an undesired "top" event or main system failure and all possible factors that could lead to the failure. Usually considered part of system safely analysis, probabilities are determined for the independent events. After simplifying the tree, the hazard evaluation team calculates the probability of the undesired event and the most likely chain of events leading to it.

Grounding, Electrostatic–conductively bonding devices to each other in order prevent build-up of dangerous electrostatic voltages. Electrostatic grounding is often required for transferring non-conducting flammable liquids from one container into another.

Hazard and operability study (HAZOP)–a hazard evaluation method for studying new or existing processes, using process flow or process instrumentation diagrams. The method is flexible and well accepted throughout industry. Excellent software and training for use is readily available. A key feature of HAZOP is the application of so-called guidewords to deviations from the design or operational intent.

Hazardous Release–an accident, incident, or event that involves an unexpected discharge or emission into the environment. The release may involve a toxic chemical, fire or explosion. A catastrophic release is a major uncontrolled emission that is defined by EPA that "presents imminent and substantial endangerment to public health and the environment." Hazardous releases may result from human errors, improper maintenance, or equipment failure in the course of industrial operations.

Ignition Temperature–sometimes called the auto-ignition temperature; this is the minimum temperature to which an object or a chemical substance must be heated in air to start and remain burning, independent of the heating source.

National Institute of Occupational Safety and Health (NIOSH)–Established under the provisions of the OSHA Act, NIOSH is the principal federal agency engaged in research, education, and training related to occupational safety and health.

Over-pressure–the pressure on an object as a result of an impacting shock wave.

Oxidizer–a chemical other than a blasting agent or explosive that initiates or promotes combustion in other materials, thereby causing fire either of itself or through the release of oxygen or other gases.

Performance Indicators–in contrast to direct measures of outcomes, performance indicators point out or suggest a means of determining achievement of established objectives. Performance indicators may be quantified by measuring the change in the level of activity (e.g. training, safety reviews, job safety analyses, etc.).

Permissible Exposure Limit (PEL)–the employee exposure to any substance listed on the MSDS shall not exceed the exposure limits specified in either milligrams per cubic meter on in parts per million. The limits are expressed either as 8-hour time weighted averages for a work shift or as acceptable ceiling concentrations experienced within the 8-hour shift. PELs are to be determined from breathing-zone air samples.

Process (OSHA definition)–any activity involving a highly hazardous chemical including any use, storage, manufacturing, handling, or the on-site movement of such chemicals, or combination of these activities. For purposes of this definition, any group of vessels, which are interconnected and separate vessels that are located such that a highly hazardous chemical could be involved in a potential release, shall be considered a single process.

Process Hazards Analysis (PHA) (OSHA definition)–a hazard evaluation on processes covered by 29 CFR 1910.119. The analysis must be appropriate to the complexity of the process and identify, evaluate and control the hazards involved in the process. Employers must determine and document the priority order for conduction process hazard analyses based on a rationale that includes such considerations as extent of the process hazards, number of potentially affected employees, age of the process and operating history of the process.

Process Safety Management–a comprehensive set of management practices required under 29 CFR 1910.119 intended to eliminate or mitigate the health and safety consequences workers face from catastrophic releases of highly hazardous chemicals. The standard lists the names and threshold quantities of the chemicals that present a potential for a catastrophic event.

Solvent–an organic liquid used to dissolve other organic materials. Examples of solvents include alcohol, benzene, methyl ethyl ketone and toluene. Water is also included under the general definition since it is commonly used to dissolve other substances.

Threshold quantity (TQ)–the amount of a highly hazardous chemical expressed in pounds and listed in OSHA's process safety or EPA's risk management planning standard. TQ is the amount of material present at any point in time in a process or storage operation that triggers the regulatory requirements. EPA's definition of TQ is less specific. However, the agency provides a basis for listing the chemicals in section 68.130 of the regulation as mandated by Congress, on the EHS list (having a vapor pressure of 10 mm mercury or greater), a toxic gas or a material with a history of accidents.

Threshold limit value (TLV)–a concentration limit for workers exposed to regulated air contaminants over a period of time, generally 8 hours, day after day with no harmful effect. Values are established by the ACGIH®. Individuals vary widely in sensitivity to chemicals and exposures below the TLV may not prevent discomfort, aggravation of a preexisting condition, or occupational illness. The limit values are derived from laboratory testing on animals, tests on humans, and/or industrial experience. Refer to the current "Guide to Occupational Exposure Values" compiled by ACGIH.

Unsafe condition–any physical state or physical, electrical, chemical or other hazard that makes an accident more likely to happen. Most of the OSHA standards describe requirements for safe work conditions. Facilities generally inspect the physical plant for the condition of slip, trip and fall hazards, chemical and physical hazards, electrical hazards, noise, fire protection and emergency exits, powered industrial trucks, personal protective equipment and ventilation and air quality. (Jack E. Daugherty, Industrial Safety Management, p. 242, Government Institutes, Rockville, Md., 1999.)

Index

3M, 15, 290, 342

Accident, Incident Investigation 89-124,
204, 223, 229, 322, 340, 344
  Corrective Action 111
  Documentation Review 106
  Facts and Causal Factors 107
  Final Report 112
  Interviews 104
  Levels of Accident Investigation 91
  Policy and Procedures 97, 98, 116
  Potential Process Incidents 116
  Staffing and Resources 96
  Time Sequencing 108, 121
Accident and Release Prevention,
  Mitigation 18, 53, 81, 85, 110, 111,
  132, 153, 186, 240, 249, 250, 257,
  258, 309, 316, 342, 357
Accountability 197, 199, 201-208, 210,
  215, 267, 307, 340, 346
Acetaldehyde 43, 255
Acetic Aldehyde 43
Acetone 16, 43, 132
Acrylonitrile 16, 254, 255, 364,
Acute Exposure 1, 4, 13, 373
Air Contaminants, Hazardous Air Pollutants
  (HAPs) 1, 15, 119, 225, 255
Air Dispersion Models 15, 29, 77, 352, 363-
371
Air Monitoring, Sampling 15, 16, 20, 29,
  70, 84, 132, 133, 141, 145, 343
Alarm System 85
Allyl Alcohol 255
ALOHA®, CAMEO® 356, 370
Alternative Scenario- See Scenarios
American Chemistry Council 315, 333
American Conference of Governmental
  Industrial Hygienists (ACGIH) 3, 13, 23,
  310, 373
American Institute of Chemical Engineers,
  (AIChE), Center for Chemical Process
  Safety (CCPS) 61, 87, 73, 75, 82, 87,
  124, 159, 201, 271, 285, 306, 315, 333

American National Standards Institute
  (ANSI) 49, 70, 73, 242
American Petroleum Institute (API) 47, 73,
  242, 315
American Society of Mechanical Engineers
  (ASME) 73, 81, 242, 276
American Society for Testing and Materials
  (ASTM) 73, 242
American Society of Safety Engineers 306,
  315
Ammonia, Anhydrous Ammonia 3, 25, 192,
  253, 255, 263, 360, 364
Amyl Acetate 45
Anheuser-Busch 331
Annual Safety and Health Plan 197, 198,
  298, 301, 347
Arsenous Trichloride 254
Asbestos 192
Audits, Assessments 214, 215, 229, 348,
  337, 338, 340, 341, 348

Behavior, Behavior Based Systems 177-186,
  300, 346
Behavior Observation Form 181
Benchmarking 80, 349
Benzene 16, 25, 263
Boiling Liquid-Expanding Vapor Explosion
  (BLEVE) 37, 373
Bonding and Grounding, see Electrostatic
  Discharge.
Boron Trifluoride 254
Bromine 253, 255
Butadiene 255
Butyl Alcohol 16, 45

Carbon Dioxide 264
Carbon Disulfide 44, 254
Carbon Monoxide 192
Case Histories, Studies 95, 118-123, 285

Catastrophe, Catastrophic Consequences,
  Release 15, 64, 75, 94, 95, 116, 143,
  167, 222, 227, 252, 373

Government Institutes Mini-Catalog

PC #	ENVIRONMENTAL TITLES	Pub Date	Price*
629	ABCs of Environmental Regulation	1998	$65
672	Book of Lists for Regulated Hazardous Substances, 9th Edition	1999	$95
4100	CFR Chemical Lists on CD ROM, 1999-2000 Edition (single user)	1999	$150
512	Clean Water Handbook, Second Edition	1996	$115
581	EH&S Auditing Made Easy	1997	$95
673	EH&S CFR Training Requirements, Fourth Edition	1999	$99
825	Environmental, Health and Safety Audits, 8th Edition	2001	$115
548	Environmental Engineering and Science	1997	$95
643	Environmental Guide to the Internet, Fourth Edition	1998	$75
820	Environmental Law Handbook, Sixteenth Edition	2001	$99
688	Environmental Health & Safety Dictionary, Seventh Edition	2000	$95
821	Environmental Statutes, 2001 Edition	2001	$115
4099	Environmental Statutes on CD ROM for Windows-Single User, 1999 Ed.	1999	$169
707	Federal Facility Environmental Compliance and Enforcement Guide	1999	$115
708	Federal Facility Environmental Management Systems	2000	$99
689	Fundamentals of Site Remediation	2000	$85
515	Industrial Environmental Management: A Practical Approach	1996	$95
510	ISO 14000: Understanding Environmental Standards	1996	$85
551	ISO 14001: An Executive Report	1996	$75
588	International Environmental Auditing	1998	$179
518	Lead Regulation Handbook	1996	$95
582	Recycling & Waste Mgmt Guide to the Internet	1997	$65
615	Risk Management Planning Handbook	1998	$105
603	Superfund Manual, 6th Edition	1997	$129
566	TSCA Handbook, Third Edition	1997	$115
534	Wetland Mitigation: Mitigation Banking and Other Strategies	1997	$95

PC #	SAFETY and HEALTH TITLES	Pub Date	Price*
697	Applied Statistics in Occupational Safety and Health	2000	$105
547	Construction Safety Handbook	1996	$95
553	Cumulative Trauma Disorders	1997	$75
663	Forklift Safety, Second Edition	1999	$85
709	Fundamentals of Occupational Safety & Health, Second Edition	2001	$69
612	HAZWOPER Incident Command	1998	$75
662	Machine Guarding Handbook	1999	$75
535	Making Sense of OSHA Compliance	1997	$75
718	OSHA's New Ergonomic Standard	2001	$95
558	PPE Made Easy	1998	$95
683	Product Safety Handbook	2001	$95
598	Project Management for EH&S Professionals	1997	$85
658	Root Cause Analysis	1999	$105
552	Safety & Health in Agriculture, Forestry and Fisheries	1997	$155
669	Safety & Health on the Internet, Third Edition	1999	$75
668	Safety Made Easy, Second Edition	1999	$75
590	Your Company Safety and Health Manual	1997	$95

Government Institutes

4 Research Place, Suite 200 • Rockville, MD 20850-3226
Tel. (301) 921-2323 • FAX (301) 921-0264
Email: giinfo@govinst.com • Internet: http://www.govinst.com

Please call our customer service department at (301) 921-2323 for a free publications catalog.

CFRs now available online. Call (301) 921-2355 for info.

* All prices are subject to change; please call for current prices.

Government Institutes Order Form

4 Research Place, Suite 200 • Rockville, MD 20850-3226
Tel (301) 921-2323 • Fax (301) 921-0264
Internet: http://www.govinst.com • E-mail: giinfo@govinst.com

4 EASY WAYS TO ORDER

1. Tel: **(301) 921-2323**
Have your credit card ready when you call.

2. Fax: **(301) 921-0264**
Fax this completed order form with your company purchase order or credit card information.

3. Mail: **Government Institutes Division**
ABS Group Inc.
P.O. Box 846304
Dallas, TX 75284-6304   USA
Mail this completed order form with a check, company purchase order, or credit card information.

4. Online: **Visit http://www.govinst.com**

PAYMENT OPTIONS

❑ **Check** *(payable in US dollars to **ABS Group Inc.** **Government Institutes Division**)*

❑ **Purchase Order** *(This order form must be attached to your company P.O. <u>Note</u>: All International orders must be prepaid.)*

❑ **Credit Card**   ❑ VISA   ❑ MasterCard   ❑

Exp. ___ /____

Credit Card No. _____

Signature _____

(Government Institutes' Federal I.D.# is 13-2695912)

CUSTOMER   INFORMATION

Ship To: (Please attach your purchase order)

Name: _____
GI Account # *(7 digits on mailing label)*: _____
Company/Institution: _____
Address: _____
(Please supply street address for UPS shipping)

City: _____ State/Province: _____
Zip/Postal Code: _____ Country: _____
Tel: (   ) _____
Fax: (   ) _____
E-mail Address: _____

Bill To: (if different from ship-to address)

Name: _____
Title/Position: _____
Company/Institution: _____
Address: _____
(Please supply street address for UPS shipping)

City: _____ State/Province: _____
Zip/Postal Code: _____ Country: _____
Tel: (   ) _____
Fax: (   ) _____
E-mail Address: _____

Qty.	Product Code	Title	Price

30 DAY MONEY-BACK GUARANTEE
If you're not completely satisfied with any product, return it undamaged within 30 days for a full and immediate refund on the price of the product.

Subtotal _____
MD Residents add 5% Sales Tax _____
Shipping and Handling (see box below) _____
Total Payment Enclosed _____

SOURCE CODE: BP03

Shipping and Handling	Sales Tax
Within U.S:	Maryland 5%
1-4 products: $6/product	Texas 8.25%
5 or more: $4/product	Virginia 4.5%
Outside U.S:	
Add $15 for each item (Global)	